高职计算机类精品教材

# Android项目开发入门

马 杰 编著

中国科学技术大学出版社

## 内 容 简 介

本书为海南软件职业技术学院课程改革成果之一,面向Android编程的初学者,以一个完整的项目案例(共20多个知识模块)为核心,以项目开发步骤组织章节顺序,对知识模块进行重点分析、论述,避免代码的简单堆砌。

本书可作为高职院校软件开发相关专业教材,也可作为相关从业人员的自学参考书。

### 图书在版编目(CIP)数据

Android项目开发入门/马杰编著.—合肥:中国科学技术大学出版社,2020.1
ISBN 978-7-312-04791-6

Ⅰ. A… Ⅱ. 马… Ⅲ. 移动终端—应用程序—程序设计—高等职业教育—教材　Ⅳ. TN929.53

中国版本图书馆CIP数据核字(2019)第207718号

出版　中国科学技术大学出版社
　　　安徽省合肥市金寨路96号,230026
　　　http://press.ustc.edu.cn
　　　https://zgkxjsdxcbs.tmall.com
印刷　合肥市宏基印刷有限公司
发行　中国科学技术大学出版社
经销　全国新华书店
开本　787 mm×1092 mm　1/16
印张　17.75
字数　444千
版次　2020年1月第1版
印次　2020年1月第1次印刷
定价　38.00元

# 前　　言

如何入门是进入软件开发领域的人面临的首个难题，Android初学者既要具备面向对象编程的基础，又要能将其与Android自身特点结合起来，才能理解Android程序的开发原理，达到入门的水平。

当前，Android的参考书往往非常"厚重"，囊括Android开发各个方面的知识，从UI界面到传感器、3D开发乃至NDK，或者提供众多项目案例，从信息管理系统到游戏软件等。这样的参考书适合用来对知识和应用进行深入学习，但是对于初学者却不见得适合，对于课堂教学也很难适用。一方面，各类知识的罗列和分析所形成的繁杂的知识点，让初学者陷入迷茫之中；另一方面，大量项目的案例代码前后关系复杂、封装层次多，让初学者无从拆解，加上很多算法和思想难以理解，导致初学者学习起来容易丧失信心。

本书试图在Android入门学习和项目案例的开发之间寻找一个切入点。知识点围绕案例来组织，以明晰、够用为原则，以培养学习兴趣、锻炼实践能力、提高学习质量为目标。案例为"个人记账本"，通过简化、分解，以由浅到深、由直观到抽象的方式组织章节内容，让初学者一步一步地实现软件的设计，获得学习的成就感。本书主要有以下一些特点：

（1）面向Android初学者，以一个项目案例为基础，以入门学习的要点和需求组织内容。

（2）清晰地描述每一个开发步骤，项目从创建到运行一步一步地演示，结合多年实践和教学经验，把初学者易出问题、难以理解的地方讲透彻。

（3）以案例"个人记账本"的开发流程组织内容，如界面设计与实现、界面跳转、欢迎功能、登录、账目存储等，让初学者能够学有所用、举一反三，也能减少大量项目案例产生的代码堆砌问题。

（4）知识点介绍力求简练实用，避免直接翻译 API 文档。例如，对于组件的属性描述，不仅介绍基本功能，还明确地介绍取值范围，给出取值示例。

（5）代码基于 Android Studio 3.2 编写。Android Studio 版本升级以后，一些新特性和细节变化使得初学者难以模仿之前的版本来创建项目，在阅读当前既有的参考书时会遇到障碍。本书所有代码基于 Android Studio 3.2 编写并作了测试运行，以便让初学者以较新的开发平台版本学习。

本书面向 Android 程序开发入门人员，特别是高校学生，可作为他们的教材或者参考书。读者学完本书的内容、培养了足够的兴趣爱好后，可再进一步深入地学习 Android 开发的知识和技能。

在本书编写的过程中，感谢陈冬明、仲超、金汋炜、刘淳滨、韩正光、李柳萍、黄辉辉、黄傲、陈文艺、刘慧、梁浒、谢鹏勤、蓝川涪在代码调试、知识点选取方面的实践和努力，感谢家人的理解和支持，特别感谢范莹、马翊睿。

因学识有限，书中不妥之处在所难免，敬请读者批评指正。希望本书的读者能静心阅读、动手实践，学有所用、学有所成。

# 目　录

前言 ……………………………………………………………………………………（ i ）

## 第1章　Android入门基础 ………………………………………………………（ 1 ）
1.1　Android简介 …………………………………………………………………（ 1 ）
1.2　第一个Android程序 …………………………………………………………（ 2 ）
1.3　Android系统架构 ……………………………………………………………（ 9 ）

## 第2章　Android UI开发 …………………………………………………………（ 11 ）
2.1　Android UI简介 ………………………………………………………………（ 11 ）
2.2　常用UI组件 …………………………………………………………………（ 11 ）
2.3　布局 …………………………………………………………………………（ 31 ）
2.4　布局嵌套 ……………………………………………………………………（ 56 ）
2.5　UI组件在Java代码和XML文件中调用 ……………………………………（ 58 ）
2.6　列表(ListView)与适配器 ……………………………………………………（ 60 ）
2.7　简洁灵活的列表(RecyclerView) ……………………………………………（ 68 ）

## 第3章　Android事件处理 ………………………………………………………（ 75 ）
3.1　基于监听器的事件处理 ……………………………………………………（ 75 ）
3.2　基于回调的事件处理、LogCat ……………………………………………（ 88 ）
3.3　基于Handler的事件处理 ……………………………………………………（ 92 ）
3.4　界面跳转 ……………………………………………………………………（ 96 ）
3.5　Activity四种启动模式 ………………………………………………………（112）
3.6　关于Context的说明 …………………………………………………………（115）

## 第4章　项目主要界面设计与实现 ………………………………………………（117）
4.1　自动跳转的欢迎页 …………………………………………………………（117）
4.2　滑屏引导页——ViewPager …………………………………………………（122）
4.3　主功能页——Fragment ……………………………………………………（128）
4.4　账目列表滑屏切换(ViewPager+Fragment) ………………………………（143）

  4.5 记账界面 ································································································ (155)
  4.6 图表统计界面 ······························································································ (156)

## 第5章 项目中的数据存取 ············································································ (161)
  5.1 引导页不再出现——Shared Preferences ······················································· (161)
  5.2 本地数据存储——SQLite数据库 ·································································· (166)
  5.3 File文件操作 ······························································································ (186)

## 第6章 使用手机相册——ContentProvider ······················································· (197)
  6.1 系统提供的ContentProvider ········································································ (197)
  6.2 自定义ContentProvider ··············································································· (203)
  6.3 主动监听ContentProvider数据变化 ······························································· (204)
  6.4 为列表项选择相册中的图片作为图标 ·································································· (205)
  6.5 为列表项拍照并添加照片作为图标 ····································································· (211)

## 第7章 背景音乐——Service与BroadcastReceiver ············································ (213)
  7.1 Service ······································································································· (213)
  7.2 音乐播放器 ································································································· (222)
  7.3 BroadcastReceiver ····················································································· (229)
  7.4 背景音乐——在Service中播放音乐 ····························································· (238)

## 第8章 动画Logo——绘图与动画 ·································································· (243)
  8.1 绘图 ············································································································ (243)
  8.2 动画 ············································································································ (248)
  8.3 跳转动画 ····································································································· (256)
  8.4 基于SurfaceView的动画 ············································································· (259)

## 第9章 手机传感器概述 ··················································································· (263)
  9.1 传感器的使用方法 ······················································································· (263)
  9.2 常用传感器 ································································································· (265)
  9.3 传感器使用示例与测试 ················································································· (267)

## 附录 ········································································································ (271)
  附录1 Android Studio下载与安装 ································································ (271)
  附录2 创建和运行第一个Android项目 ······························································ (272)
  附录3 合理使用包（package）管理项目目录 ···················································· (276)
  附录4 Android Studio常用设置 ·································································· (277)

# 第1章　Android入门基础

## 1.1 Android简介

Android是一种基于Linux的开源的操作系统,主要用于移动设备,如智能手机、平板电脑、可穿戴设备、智能电子设备等。目前,Android是全世界范围内使用最广的移动操作系统,基于Android的各类应用软件需求旺盛,Android已经成为移动互联网领域最大的应用平台。

### 1.1.1 Android发展简史

2003年10月,安迪·鲁宾创建了Android公司,开发出一款对所有软件开发者开放的移动平台。

2005年8月,Google公司收购了Android公司及其团队并继续推进Android项目的开发。

2007年11月,Google公司正式对外展示了Android的操作系统,并宣布成立开放手持设备联盟(Open Handset Alliance)。开放手持设备联盟由全球顶尖的手机制造商、软件开发商、电信运营商以及芯片制造商组成,初始会员包括三星电子、摩托罗拉、谷歌、中国移动、英特尔、高通公司、德州仪器等65家商业公司,其宗旨是以开源的形式一同研发、改良和推广Android系统。

2008年9月,由HTC公司定制的首款Android手机 T-Mobile G1发布,其搭载了Android 1.0系统。

2011年8月,Android手机占据全球智能手机市场48%的份额,超过Symbian(塞班)系统,跃居全球第一。

2016年12月,Strategy Analytics公司发布调研报告称Android手机的全球市场份额达到了87.5%,iOS(苹果)系统的全球市场份额为12.1%。

2018年6月,市场研究公司IDC发布报告称,2018年Android手机销量在全球市场的占比约为85%,未来五年其销量有望以2.4%的年复合增长率保持增长。同时,网站通信流量监测机构StatCounter公布的数据显示,Android以41.66%的占比位列全球操作系统第一,排名第二的是Windows操作系统,其占比为35.93%。

### 1.1.2 Android 版本及应用分布

自 Android 1.0 发布以来，Android 版本更新速度很快，市面上应用的 Android 版本较多，在为用户提供更优服务性能的同时，也导致了碎片化问题，不同版本间性能的变化对应用开发中的版本适配带来了一定的影响。根据 Google 公司发布的数据，截止到 2018 年 7 月，各个 Android 版本的使用分布情况如表 1.1 所示。

表 1.1 Android 各版本使用分布情况

| 版本名称 | 版本号 | API | 占比 |
| --- | --- | --- | --- |
| frogy(冻酸奶) | 2.2.x | 8 | 0% |
| gingerbread(姜饼) | 2.3.3—2.3.7 | 10 | 0.2% |
| Ice Cream Sandwich(冰激凌三明治) | 4.0.3—4.0.4 | 15 | 0.3% |
| Jelly Bean(果冻豆) | 4.1.x | 16 | 1.2% |
|  | 4.2.x | 17 | 1.9% |
|  | 4.3 | 18 | 0.5% |
| KitKat(奇巧巧克力) | 4.4 | 19 | 9.1% |
| Lollipop(棒棒糖) | 5.0 | 21 | 4.2% |
|  | 5.1 | 22 | 16.2% |
| Marshmallow(棉花糖) | 6.0 | 23 | 23.5% |
| Nougat(牛轧糖) | 7.0 | 24 | 21.2% |
|  | 7.1 | 25 | 9.6% |
| Oreo(奥利奥) | 8.0 | 26 | 10.1% |
|  | 8.1 | 27 | 2.0% |

## 1.2 第一个 Android 程序

### 1.2.1 Hello World 程序分析

Android 应用开发基于 Java 语言，但是为了适应移动开发的需要，Android SDK 与 Java SDK 并不完全相同。Android SDK 引用了大部分的 Java SDK，同时还自定义了很多开发工具包，这使得 Android 在程序框架、界面设计、网络访问、数据存储等很多方面与 Java 有着明显区别。

本书以 Google 推荐的 Android Studio 作为开发环境，Android Studio 基于 Gradle 进行项目构建。关于 Android Studio 开发环境的安装配置、Android 的项目创建和程序运行步骤，请参见本书附录。

**1. 程序执行**

在 Hello World 项目的"Android"视图模式下，打开 manifests 目录中的 AndroidManifest.xml 程序清单文件，如下：

```xml
<?xml version="1.0" encoding="utf-8"?>
<manifest xmlns:android="http://schemas.android.com/apk/res/android"
    package="cc.turbosnail.helloworld">
    <application
        android:allowBackup="true"
        android:icon="@mipmap/ic_launcher"
        android:label="@string/app_name"
        android:roundIcon="@mipmap/ic_launcher_round"
        android:supportsRtl="true"
        android:theme="@style/AppTheme">
        <activity android:name=".MainActivity">
            <intent-filter>
                <action android:name="android.intent.action.MAIN" />
                <category android:name="android.intent.category.LAUNCHER" />
            </intent-filter>
        </activity>
    </application>
</manifest>
```

程序启动时,会先读取AndroidManifest.xml清单文件中描述的信息。<activity>标记中的android:name属性指定了当前类是MainActivity,即"MainActivity.java",<activity>内部子元素<intent-filter>中<action android:name="android.intent.action.MAIN"/>语句则声明了该类为MAIN,即程序执行的入口,所以当前项目运行时将根据清单文件的说明以"MainActivity.java"为执行起点。

关于AndroidManifest.xml文件,进一步说明如下:

(1) <manifest>标签。

☞ xmlns:android=http://schemas.android.com/apk/res/android

命名空间声明,当前的命名空间为Android的默认命名空间,它提供了各种Android的标准元素和属性。

☞ package="cc.turbosnail.helloworld"

声明项目的包名。

(2) <application>标签。

一个AndroidManifest.xml中必须含有一个application标签,这个标签声明应用程序的组件及其属性。

☞ android:allowBackup="true"

是否允许通过adb命令备份和恢复应用程序数据,取值为"true"或者"false"。

☞ android:icon="@mipmap/ic_launcher"

声明APP在屏幕上的程序图标,项目安装后,屏幕上将出现该图标作为程序图标,用户可以自定义该图标。图标文件的路径用"@"的形式说明,存放路径为项目目录res/mipmap/,

图片格式推荐使用 png 格式。默认的图标是 ic_launcher，"@mipmap/ic_launcher" 表示 mipmap 文件夹下的 ic_launcher.png 文件。

☞ android:label="HelloWorld"

声明 APP 的标题名，程序运行时显示在最上方的标题栏中，用户可以自定义 APP 标题字符串。

☞ android:roundIcon="@mipmap/ic_launcher_round"

声明 APP 在屏幕上的程序图标为圆形图标样式，用户可以自定义该图标。图标文件存放路径为项目目录 res/mipmap。这是 Android 8.0 开始提供的一种让程序图标不局限于圆角矩形样式的图标适配解决方案。

☞ android:supportsRtl="true"

声明是否支持"从右到左"显示风格，取值为"true"或者"false"。如果设置为"true"，则在手机的开发人员选项中选择"强制使用从右到左的布局方向"后，程序将呈现从右到左的显示风格。

☞ android:theme="@style/AppTheme"

声明当前 APP 的主题，允许用户使用自定义的主题。打开项目目录中的 style.xml，可以看到当前使用的主题"AppTheme"，它继承了 Theme.AppCompat.Light.DarkActionBar 样式。

```
<resources>
    <!-- Base application theme. -->
    <style name="AppTheme" parent="Theme.AppCompat.Light.DarkActionBar">
        <!-- Customize your theme here. -->
        <item name="colorPrimary">@color/colorPrimary</item>
        <item name="colorPrimaryDark">@color/colorPrimaryDark</item>
        <item name="colorAccent">@color/colorAccent</item>
    </style>
</resources>
```

（3）<activity>标签。

activity 元素作为 application 的子元素，用来声明一个活动的 Activity。所有的活动 Activity 都必须通过 activity 元素进行注册声明，否则它不会被系统看到，也不会被运行。

android:name=".MainActivity"。android:name 属性采用类名的简写方式声明 Activity 类，格式为".ClassName"，不用添加".java"扩展名，类名前的"."表示当前包（package）。如果当前类在子包 test 中，则应写成 android:name=".test.MainActivity"。

（4）<intent-filter>标签。

intent-filter 标签用来定义一个过滤器，对其父元素的功能进行声明。它一般包含 action、category、data 三个子元素。

（5）<action>标签。

action 标签只有一个 name 属性，常见的 android:name 值为 "android.intent.action.MAIN"，声明此 Activity 作为应用程序的执行入口。

(6)<category>标签。

category只有name属性。常见的android:name值为android.intent.category.LAUNCHER，声明当前Activity应用程序将显示在桌面程序列表中。

AndroidManifest.xml是Android项目必须的清单文件，除了有以上标签外，还有大量用来进行组件注册、访问授权、API级别声明等功能的标签，它是Android基于XML格式的核心清单文件。

2. 执行MainActivity.java

AndroidManifest.xml文件中声明了程序执行入口是MainActivity，在项目目录java/cc.turbosnail.helloworld中打开MainActivity.java文件，看到自动生成的代码如下：

```java
package cc.turbosnail.helloworld;
import android.support.v7.app.AppCompatActivity;
import android.os.Bundle;
public class MainActivity extends AppCompatActivity {
    @Override
    protected void onCreate(Bundle savedInstanceState) {
        super.onCreate(savedInstanceState);
        setContentView(R.layout.activity_main);
    }
}
```

MainActivity是AppCompatActivity的子类，AppCompatActivity是Activity类经过多级继承后的子类，Activity是Android中用来负责与用户交互的组件，一个Activity可以理解为一个覆盖整个屏幕的界面容器，用户界面上的内容均需要放置在Activity中显示。AppCompatActivity除了提供基本的容器功能外，还提供标题栏、样式等更丰富的应用支持。

MainActivity运行时，onCreate方法会被调用，表示创建一个Activity对象时执行该方法，Bundle类型的参数savedInstanceState用来以"key-value"键值对的形式存储当前Activity的状态值。

setContentView(R.layout.ly_main)语句可翻译为：设置当前显示的视图为R.layout.ly_main。

整段代码的含义可以理解为：程序启动时，MainActivity的onCreate方法被调用，该方法通过执行setContentView语句将R.layout.ly_main设置成当前的显示内容。

3. R.layout.ly_main

把HelloWorld项目的打开方式由默认的"Android"切换到"Packages"模式，如图1.1所示。

图1.1 Package模式

在HelloWorld项目的"Packages"视图模式下,打开app/cc.turbosnail.helloworld/目录,可以看到一个R文件,打开该文件,然后在当前代码中搜索"activity_main"(可以通过"Ctrl+F"键打开搜索栏),能看到如下代码:

```
public static final class layout {
    public static final int activity_main=0x7f09001c;
}
```

从层次关系上可以看出这里的activity_main即为R.layout.activity_main。

R文件用来对所有的外部资源进行注册,并自动地生成一个int类型的唯一标识。外部资源一般在res目录中存放,代码中的R.layout.activity_main即对应res/layout目录中的activity_main.xml文件。

R文件中有attr、drawable、layout、string等多个静态内部类,每个静态内部类对应一种资源文件类型,其中的每个静态常量代表一个外部资源文件的标示。R文件头部有注释文字"自动生成文件,请不要修改",R文件会自动对添加到项目中的外部资源进行注册,开发者不能手动修改R文件中的信息。

### 4. activity_main.xml文件

切换HelloWorld项目的视图模式到"Android"模式下,打开res/layout目录中的activity_main.xml文件,点击代码编辑框下方的"Text"标签切换到文本视图模式,可以看到如下代码:

```
<?xml version="1.0" encoding="utf-8"?>
<android.support.constraint.ConstraintLayout xmlns:android="http://schemas.android.com/apk/res/android"
    xmlns:app="http://schemas.android.com/apk/res-auto"
    xmlns:tools="http://schemas.android.com/tools"
    android:layout_width="match_parent"
```

```xml
    android:layout_height="match_parent"
    tools:context=".MainActivity">
    <TextView
        android:layout_width="wrap_content"
        android:layout_height="wrap_content"
        android:text="Hello World!"
        app:layout_constraintBottom_toBottomOf="parent"
        app:layout_constraintLeft_toLeftOf="parent"
        app:layout_constraintRight_toRightOf="parent"
        app:layout_constraintTop_toTopOf="parent" />
</android.support.constraint.ConstraintLayout>
```

这是基于XML文件形式实现的程序界面。<android.support.constraint.ConstraintLayout>标签代表的是一种被称为约束布局的布局方式，布局作为一种容器用来放置组件，同时还根据自身的特点约定组件的排列方式。

<TextView>是文本标签组件，可以用来显示文本信息，android:text="Hello World!"表示当前显示的文本信息是"Hello World!"。

android:layout_width="wrap_content"和android:layout_height="wrap_content"这两句用来设置组件的宽度和高度，"wrap_content"是常量之一，表示根据自身内容自适应大小。

app:layout_constraintBottom_toBottomOf="parent"等语句用来在当前布局下为组件设置约束位置，例如本句的作用是组件底部与它父容器底部建立约束关系。

Android允许开发者基于XML资源文件的形式进行界面设计，这样可以有效地实现界面设计和逻辑处理的分离。它提供了多种布局样式和组件，还允许自定义布局和组件，以方便开发者进行灵活的界面设计。

经过分析，我们可以看到Android程序的执行流程如下：
（1）读取AndroidManifest.xml文件，找到启动类MainActivity。
（2）执行MainActivity中的onCreate方法。
（3）执行setContentView()方法，设置当前界面为R.layout.activity_main；R.layout.activity_main代表layout/activity_main.xml文件，此为界面显示文件。

### 1.2.2　Android Studio项目目录

除了上一节已经介绍的项目目录和文件，在项目的"Android"视图模式下下逐一展开各个节点，可以看到项目目录结构，如图1.2所示。

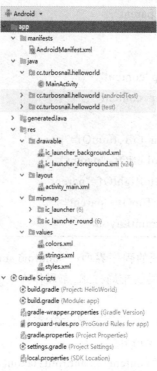

图1.2　Android Studio项目目录

在编写Android程序的过程中,推荐使用"Android"视图模式,对Java代码、资源文件、配置文件、构建文件进行统一管理,在该视图模式下,目录中各个文件的作用介绍如表1.2所示。

表1.2　Android Studio项目目录功能

| 目录名称 | 存放内容 |
| --- | --- |
| manifests | 存放清单文件AndroidManifest.xml |
| java | 存放Java源代码和测试代码 |
| generatedJava | 存放项目生成的Java代码 |
| res | 存放项目资源文件 |
| -drawable | 存储图片文件、绘图文件等 |
| -layout | 存放布局文件 |
| -mipmap | 存放图标文件,自动根据屏幕分辨率进行选择,当前5个图标的分辨率分别是48×48(mdpi)、72×72(hdpi)、96×96(xhdpi)、144×144(xxhdpi)、192×192(xxxhdpi),还有一个是基于前景/背景自定义图片外观的XML文件 |
| -values | 存放需要引用值的资源 |
| --colors.xml | 存放自定义的颜色列表 |
| --strings.xml | 存放字符串,把字符串作为资源文件定义然后引用 |
| --styles.xml | 存放样式文件,对界面或者组件的整体样式进行控制 |
| Gradle Scripts | Android Studio基于Gradle进行构建,Gradle是一个基于Apache Ant和Apache Maven概念的项目自动化建构工具 |

## 1.3 Android系统架构

Android系统采用分层架构,从底层到应用程序分为Linux内核(Linux Kernel)、库及Android运行时环境(Libraries&Android Runtime)、应用程序框架(Application Framework)、应用层(Applications),其架构模型如图1.3所示。

### 1. Linux内核

Android系统基于Linux内核开发,Linux内核层为系统提供了硬件的驱动程序、网络、电源、系统安全、内存管理等功能。它包含了显示驱动、相机驱动、蓝牙驱动、Flash内存驱动、IPC驱动、USB驱动、键盘驱动、WiFi驱动、音频驱动、电源管理等一系列组件。

### 2. 库及Android运行时环境

Libraries提供系统核心库,例如用于界面管理的Surface Manager库、用于多媒体处理的MediaFramework框架、用于数据存储的SQLite库、用于3D绘图的OpenGL ES库、用于位图和矢量字体的Free Type库、用于Web浏览器的WebKit库、用于2D图形的SGL库、用于数据通信安全支持的SSL(Secure Sockets Layer,安全套接层)库、用于调用Linux系统功能的Libc库等。

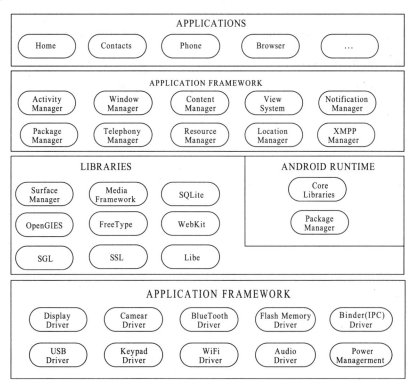

图1.3 Android系统架构

Android Runtime 包括了一个 Core Libraries(核心库)和 Dalvik Virtual Machine(Dalvik 虚拟机)。

（1）核心库提供了Java语言核心库的大多数功能。

（2）Dalvik虚拟机是用于Android的虚拟机，它基于寄存器架构，运行经过编译生成的.dex文件。Android代码基于Java语言实现，源代码为.java文件，编译成.class文件后，会进一步地针对小内存优化生成.dex文件或再进一步优化性能生成.odex文件。每一个Android应用程序都在它自己的进程中运行，都拥有一个独立的Dalvik虚拟机实例，即每个Android应用程序都拥有一个自己的Dalvik虚拟机进程。Android的执行文件打包为.apk格式发布，.apk文件是由.dex文件、资源和配置文件组成的压缩包。

从Android 4.4开始，Android系统提供了ART虚拟机。ART虚拟机的一个重要特性是AOT(ahead of time 提前编译)，在安装.apk文件的时候就将.dex直接处理成可供ART虚拟机使用的机器码文件，将.dex文件转换成可直接运行的.oat文件，每次虚拟机启动时直接执行机器码文件，这样运行效率更高。对应地，Dalvik虚拟机拥有的是JIT(Just In Time，即时编译)特性，需要每次虚拟机启动时才将.dex文件转换成机器码。

3. 应用程序框架(Application Framework)

应用程序框架提供Android开发的基础类库供程序开发人员使用。它包括了用于管理Android的最基本组件Activity Manager、用于界面窗口管理的Window Manager、用于多个程序间数据共享的Content Providers、用于界面设计的组件集合View System、用于状态栏短信或者来电等通知消息管理的Notification Manager、用于程序包管理的Package Manager、用于进行手机基本服务管理的Telephony Manager、用于资源文件管理的Resource Manager、用于GPS等位置服务的Location Manager、用于通信服务的XMPP Service等。

4. 应用层(Applications)

基于Java语言实现的Android应用程序，例如Home功能、手机通信录、拨号程序、浏览器以及第三方开发者设计实现的各类应用程序。

# 第2章 Android UI开发

## 2.1 Android UI 简介

Android UI(User Interface,用户界面)组件由 View 和 ViewGroup 组成。View 是用于界面显示和用户交互的视图对象,常用界面 UI 组件如 TextView、Button、ImageView、CheckBox 等都是 View 类的直接或者间接子类。ViewGroup 是用于容纳多个 View 或者 ViewGroup 对象的容器类组件,它本身也是 View 的子类,常用容器如 LinearLayout、RelativeLayout、FrameLayout 等布局类都是 ViewGroup 的子类或者间接子类。在 UI 开发中可以直接使用 View 类和 ViewGroup 类,也可以重写相关方法自定义组件和容器类。一个由 View 和 ViewGroup 组成的典型界面结构如图 2.1 所示。

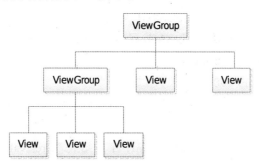

图 2.1 Android 界面组成结构示意图

## 2.2 常用 UI 组件

### 2.2.1 UI 组件常用公共属性

在界面开发中,每种 UI 组件都有自身的属性,通过为属性设置对应的值,可以让组件呈现出预期的效果,以满足大小、颜色、背景、透明度等要求。例如"android:layout_width"用来设置组件的宽度。组件一般具有自身特有的属性,也有一些公共的属性可适用于多数组件。

### 1. id 属性

为当前的 UI 组件定义一个唯一的 id 名称。该名称会在 R.java 文件中自动注册,用于在 Java 代码或者 XML 文件中对该组件的引用。id 属性定义的格式为 android:id="@+id/name",name 为开发者自定义的 id 名称,应是以字母或者下划线开头的合法标识符,如 android:id="@+id/tv_hello"。

### 2. 大小设置

一般组件都需要设置自己的宽度和高度,其 XML 属性为 android:layout_width、android:layout_height,可取值为:

- ☞ "match_parent":填充满该组件的父容器,即与父容器相同大小。
- ☞ "wrap_content":根据组件自身的内容自动调整大小。
- ☞ 像素值:用具体的像素值来设置大小,推荐使用单位 dp。

例如:

android:layout_width="match_parent"

android:layout_width="100dp"

Android 的像素单位包括 dp(设备独立像素)、px(绝对像素)、pt(磅)、sp(放大像素)、in(英寸)、mm(毫米)等。在开发中推荐使用 dp 作为长度单位,dp 可以让 UI 组件在不同分辨率的屏幕上按比例保持相近的外观效果。在开发中推荐使用 sp 作为字体大小的单位,sp 与 dp 类似,但它接受用户在手机设置中自定义的字体大小设置。

### 3. 颜色设置

Android 接受用六位 16 进制数表示 RGB 颜色的方式,例如:黑色为"#000000"、白色为"#FFFFFF"。常用需要设置颜色的 XML 属性有 android:textColor、android:background、android:textColorHighlight、android:textColorHint 等。例如:

android:textColor="#FF0000"

在应用开发中,可以自定义一个基于 XML 的颜色文件,既方便调用,也可以用来统一软件的颜色风格。具体做法为:

(1) 打开 res/values 目录下的 colors.xml,目录结构为 res/values/colors.xml,可以看到 Android 项目自带的一个颜色自定义文件,内容为:

```xml
<?xml version="1.0" encoding="utf-8"?>
<resources>
    <color name="colorPrimary">#008577</color>
    <color name="colorPrimaryDark">#00574B</color>
    <color name="colorAccent">#D81B60</color>
</resources>
```

(2) 参照这个结构,在 colors.xml 的 <resources> 标签中添加自定义的颜色。例如:

```
<color name="white">#FFFFFF</color><!--白色-->
<color name="orange">#FFA500</color><!--橙色-->
<color name="purple">#800080</color><!--紫色-->
<color name="lavender">#E6E6FA</color><!--淡紫色-->
```

(3)在为组件设置颜色属性时,通过"@color/name"的形式使用自定义的颜色,例如:android:textColor="@color/lavender"

4. 使用图片

Android支持的图片格式有png、jpg、gif等位图,或者是9-Patch图片、XML描述的图形和图像对象。使用图片资源时,可以按照如下步骤将图片添加到drawable目录下。

(1)复制准备好的图片。

(2)选中drawable目录,然后点击鼠标右键,选择"Paste",如图2.2所示。

(3)在弹出的对话框中可以修改图片的名字,请使用合法的、有明确意义的字符串命名资源文件,如图2.3所示。点击"OK"按钮后,可以在当前项目的res/drawable目录下看到新添加的图片资源文件。

图2.2 复制图片到项目目录　　　　　图2.3 修改图片名称

(4)通过"@drawable/name"使用图片资源,如为当前组件设置背景属性时可以使用刚才添加的图片资源"bg_welcome.png"如下:

android:background="@drawable/bg_welcome"

由于不同型号手机的屏幕分辨率、尺寸都不统一,使得物理DPI(Dots Per Inch,每英寸长度上像素点个数)出现多样,这直接影响到了程序的显示效果。为了减少图片适配产生的问题,Android项目中提供多个图片资源文件夹,用来放置同一张图片多个大小不同的版本,程序运行时会根据手机屏幕规格自动寻找分辨率适合的图片资源文件夹,读取其中的图片。例如:在当前的项目目录中,mipmap目录下的ic_launcher.png就有hdpi、mdpi、xhdpi、xxhdpi、xxxhdpi五个版本,分别用来适配大小不一的各种手机屏幕。mipmap目录一般用来存放启动时的图标文件,在存放图标时,对应的像素如表2.1所示。

表2.1 drawable图标分辨率

| drawable文件夹 | 图标分辨率 |
| --- | --- |
| drawable-mdpi | 48×48 px |
| drawable-hdpi | 72×72 px |
| drawable-xhdpi | 96×96 px |
| drawalbe-xxhdpi | 144×144 px |
| drawable-xxxhdpi | 192×192 px |

### 5. 组件的内外边距设置

一个 UI 组件在屏幕上所占的位置由外边距（margin）、边框（border）、内边距（padding）、内容（content）四部分组成，组件的宽度（width）和高度（height）指的是内容（content）的宽度和高度，内边距是内容至边框之间的部分，边框是内外边距的分界线，外边距是边框至组件位置边缘之间的部分，示意图如图 2.4 所示。

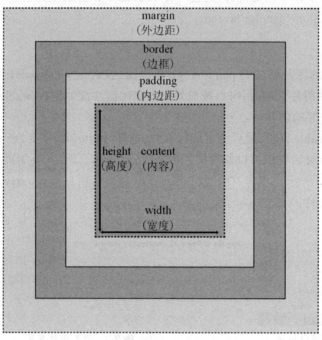

图 2.4　组件边距示意图

调整组件内边距的值可以使组件内容部分在组件边框内发生位移，调整组件外边距的值可以使组件占据的位置发生变化，即相对于它旁边的组件或者父容器的位置发生变化，调整组件边框的属性值可以改变边框的显示效果。在 XML 中调整对应属性的值如表 2.2 所示。

表 2.2　组件内外边距属性设置

| 属性声明 | 功能描述 | 属性取值 |
| --- | --- | --- |
| android:layout_marginTop | 设置上外边距 | |
| android:layout_marginBottom | 设置下外边距 | |
| android:layout_marginLeft | 设置左外边距 | |
| android:layout_marginRight | 设置右外边距 | 边距值，单位为 dp，如 20 dp |
| android:paddingTop | 设置上内边距 | |
| android:paddingBottom | 设置下内边距 | |
| android:paddingLeft | 设置左内边距 | |
| android:paddingRight | 设置右内边距 | |
| android:padding | 一次设置四个方向内边距的值 | 边距值，单位为 dp，上下左右内边距均为这个值 |
| android:layout_margin | 一次设置四个方向外边距的值 | |

## 2.2.2 文本标签(TextView)

文本标签用于在界面上显示文本信息,它可以显示单行文字,也可以显示多行文字,还可以显示带图像的文本。文本标签常用的属性如表2.3所示。

表2.3 文本标签常用属性列表

| 属性声明 | 功能描述 | 属性取值 |
| --- | --- | --- |
| android:text | 指定文本标签显示的文字 | 文本字符串,如"Hello World!" |
| android:textSize | 设置字体大小 | 数值,单位为sp |
| android:textColor | 设置文字的颜色 | 十六进制的颜色值,或者"@color/name"形式引用颜色资源 |
| android:textStyle | 设置文字样式 | "bold""italic""normal" |
| android:gravity | 设置文本内容的对齐方式 | "center""left""right""top""bottom"等,多个用"\|"隔开,如"right \| bottom" |
| android:width | 设置文本内容的宽度 | 数值,单位为dp |
| android:height | 设置文本内容的高度 | 数值,单位为dp |
| android:autoLink | 把文本中包含的特定格式显示为超链接样式,单击时,会调用默认打开方式 | "all""email""map""none""phone""web",多个用"\|"隔开,如"email \| phone" |
| android:drawableTop android:drawableRight android:drawableBottom android:drawableLeft | 在文本标签内部文本的上、右、下、左位置放置图片 | "@drawable/name",drawable目录下的图片资源 |
| android:textIsSelectable | 设置文字是否可被选中 | "true"或"false" |
| android:textColorHighlight | 设置文字被选中后的背景颜色 | 十六进制的颜色值,或者"@color/name"的形式引用颜色资源文件值 |
| android:lineSpacingExtra | 设置行间距数值 | 数值,单位为dp |
| android:lineSpacingMultiplier | 设置行间距倍数值 | 数值,无单位,如"1.5" |
| android:maxLines | 设置最大显示行数,配合android:ellipsize="end"可以在结尾将超出部分显示为省略号"…" | 整数,如"3" |
| android:lines | 设置固定的显示行数 | 整数,如"5" |
| android:ellipsize | 文字长度超出一行显示宽度时,超出部分显示省略号的位置 | "start"省略号显示在开头,"end"省略号显示在结尾,"middle"省略号显示在中间,"emarquee"获得焦点横向滚动显示 |

【例2.1】在 Android Studio 中新建一个项目,项目名为"UIDemo",打开 res/layout/activity_main.xml 文件,修改 TextView 标签的属性如下:

```xml
<TextView
    android:layout_width="wrap_content"
    android:layout_height="wrap_content"
    android:text="欢迎来到Android世界!"
    android:textSize="20sp"
    android:textColor="#FF0000"
    android:textStyle="bold"
    android:gravity="center"
    app:layout_constraintBottom_toBottomOf="parent"
    app:layout_constraintLeft_toLeftOf="parent"
    app:layout_constraintRight_toRightOf="parent"
    app:layout_constraintTop_toTopOf="parent"/>
```

运行程序,可看到结果如图 2.5 所示。

【例2.2】实现一个文字可选中、行间距 1.5 倍的文本标签。在项目"UIDemo"中,打开 res/layout/activity_main.xml 文件,修改 TextView 标签的属性如下:

```xml
<TextView
    android:layout_width="wrap_content"
    android:layout_height="wrap_content"
    android:text="大胆地假设、小心地求证。要大胆地提出假设,但这种假设还得想法子证明。所以小心地求证,想法子证实假设或者否定假设,比大胆地假设还重要。"
    android:textSize="20sp"
    android:gravity="top|left"
    android:lineSpacingMultiplier="1.5"
    android:textIsSelectable="true"
    android:textColorHighlight="#317ef3"
    app:layout_constraintBottom_toBottomOf="parent"
    app:layout_constraintLeft_toLeftOf="parent"
    app:layout_constraintRight_toRightOf="parent"
    app:layout_constraintTop_toTopOf="parent"/>
```

运行程序,可看到结果如图 2.6 所示。

第 2 章　Android UI 开发

图 2.5　文本标签属性设置　　　2.6　文字可选中的文本标签

### 2.2.3　按钮(Button)

按钮是事件编程中常用的组件,在界面设计中,可以通过背景设置、样式定义等方式实现多种外观效果的按钮。Button继承于TextView,可以使用TextView中的多数属性,在应用开发中,按钮常用的属性如表2.4所示。

表 2.4　按钮常用属性列表

| 属性声明 | 功能描述 | 属性值 |
| --- | --- | --- |
| android:text | 设置按钮上显示出来的标题 | 字符串,如"确定" |
| android:background | 设置背景颜色或图片 | 十六进制颜色值,或者"@color/name"形式引用颜色资源文件值 |
| android:drawableTop | | |
| android:drawableRight | 在按钮内部上、右、下、左位置放置图片,给按钮的标题加上小图标 | "@drawable/name",drawable 目录下的图片资源 |
| android:drawableBottom | | |
| android:drawableLeft | | |

【例2.3】创建一个按钮。在项目"UIDemo"中,首先打开 layout/activity_main.xml 文件,用"<!-- 代码段 -->"注释掉TextView标记部分,然后添加如下代码:

```
<Button
    android:layout_width="wrap_content"
    android:layout_height="wrap_content"
    android:text="这是一个按钮"/>
```

运行程序,可看到结果如图2.7所示。

图 2.7 简单按钮

【例 2.4】创建一个圆角按钮,且用户点击前后按钮显示不同颜色。

(1) 创建两个 shape 文件。

在项目"UIDemo"的 res 目录上点击右键,选择"New"→"Android resource file",如图 2.8 所示。

图 2.8 创建 resource 文件

在弹出的对话框中输入文件名、类型、根元素名称、目录名称等内容,如图 2.9 所示。

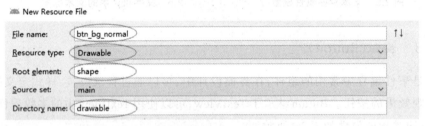

图 2.9 创建 shape 文件

点击"OK"按钮,可以看到在 res/drawable 目录中新创建的"btn_bg_normal.xml"文件,在根元素 shape 中添加如下代码:

```xml
<?xml version="1.0" encoding="utf-8"?>
<shape xmlns:android="http://schemas.android.com/apk/res/android">
    <!-- 矩形的圆角弧度 -->
    <corners android:radius="10dp" />
    <!-- 矩形的填充颜色,深蓝色 -->
    <solid android:color="#317EF3" />
</shape>
```

同理,创建第二个 shape 文件 btn_bg_pressed.xml,代码如下:

```xml
<?xml version="1.0" encoding="utf-8"?>
<shape xmlns:android="http://schemas.android.com/apk/res/android">
    <!-- 矩形的圆角弧度 -->
    <corners android:radius="10dp" />
    <!-- 矩形的填充色,浅蓝色-->
    <solid android:color="#8eb9f5" />
</shape>
```

（2）创建selector文件。

参照第一个步骤，在res目录上右键点击，创建selector.xml文件，根元素名为"selector"，如图2.10所示。

图2.10 创建selector文件

修改selector.xml的代码内容如下：

```xml
<?xml version="1.0" encoding="utf-8"?>
<selector xmlns:android="http://schemas.android.com/apk/res/android">
    <!-- Button正常状态下的背景 -->
    <item android:drawable="@drawable/btn_bg_normal" android:state_pressed="false"/>
    <!-- Button按下时的背景 -->
    <item android:drawable="@drawable/btn_bg_pressed" android:state_pressed="true"/>
</selector>
```

（3）在layout/activity_main.xml布局文件中用"<!-- 代码段 -->"注释掉上例中的<Button>标签，重新创建一个按钮，定制外观并引用selector.xml文件作为背景。

```xml
<Button
    android:layout_width="match_parent"
    android:layout_height="wrap_content"
    android:text="登录"
    android:textSize="20sp"
    android:textColor="#FFFFFF"
    android:background="@drawable/selector"/>
```

运行程序，可看到初始状态和点击状态的运行效果分别如图2.11所示。

图2.11 圆角按钮

### 2.2.4 \<selector>和\<shape>元素

在界面设计中，\<selector>元素可以方便地根据UI组件的状态改变显示效果，例如背景颜色、图片等，减少通过程序代码修改外观效果的工作量，是UI设计中比较常用的一个元素。它可以限制的列表内容如下：

```xml
<?xml version="1.0" encoding="utf-8"?>
<selector xmlns:android="http://schemas.android.com/apk/res/android" >
    <item
        android:state_pressed=["true" | "false"]//是否触摸
        android:state_focused=["true" | "false"]//是否获得焦点
        android:state_selected=["true" | "false"]//是否被状态
        android:state_checkable=["true" | "false"]//是否可选
        android:state_checked=["true" | "false"]//是否选中
        android:state_enabled=["true" | "false"]//是否可用
        android:state_window_focused=["true" | "false"] />//是否窗口聚焦
</selector>
```

使用\<shape>元素可以方便地绘制图形，相对于png等图片资源来说，它可以有效减少安装包的大小，而且能够更好地适配不同的手机。

在\<shape>标签中，可以定义的类型包括矩形（rectangle）、椭圆（oval）、线（line）、圆环（ring）等，通过android:shape属性来进行声明：

android:shape=["rectangle" | "oval" | "line" | "ring"]

\<shape>标签内部的元素用来进行详细的外观设置，主要包括以下元素：

- \<size>：设置大小，android:width、android:height分别设置宽度和高度，单位是dp。
- \<solid>：填充颜色，android:color设置颜色。
- \<corners>：圆角大小，android:radius设置圆角、数值类型，dp为单位，如"android:radius=10dp"。
- \<padding>：设置内边距。
- \<stroke>：设置边框，android:width设置边框大小，单位为dp。android:color设置边框颜色。
- \<gradient>：设置渐变色，android:startColor设置渐变起始颜色，android:endColor设置渐变结束颜色。android:angle设置渐变角度："0"：左到右，"90"：下到上，"180"：右到左，"270"：上到下。

例如，定义一个颜色渐变的圆形bg_circle.xml文件如下：

```xml
<?xml version="1.0" encoding="utf-8"?>
<shape xmlns:android="http://schemas.android.com/apk/res/android" android:shape="oval">
    <size android:height="100dp"
```

```
        android:width="100dp"/>
    <gradient
        android:startColor="#FF0000"
        android:endColor="#0000FF"
        android:angle="90"/>
</shape>
```

把它作为一个组件的背景或者图像源使用,方式分别如下:
android:background="@android:drawable/bg_circle"
android:src="@android:drawable/bg_circle"
运行效果如图 2.12 所示。

图 2.12　渐变圆形样式

<selector>和<shape>搭配使用,可以实现丰富的 UI 界面效果,在软件开发中具有广泛的应用场景。

### 2.2.5　编辑框(EditText)

编辑框允许用户在其中进行文字输入,它继承了文本标签(TextView)的多数外观属性,同时定义了与用户输入相关的属性。编辑框常用的属性如表 2.5 所示。

表 2.5　编辑框常用属性列表

| 属性声明 | 功能描述 | 属性值 |
| --- | --- | --- |
| android:inputType | 限定输入类型 | "number"数字<br>"numberDecimal"带小数点浮点数<br>"textPassword"密码<br>"textVisiblePassword"密码可见<br>"phone"电话号码<br>"textEmailAddress"电子邮件<br>"textCapWords"单词首字母大写<br>"datetime"日期时间<br>"textMultiLine"允许多行输入 |
| android:hint | 显示默认提示语,获得控制焦点后不显示 | 字符串,如"请输入用户名" |
| android:maxLength | 设置可输入文字的最大长度,超出后不允许输入 | 数字,如"12" |

【例2.5】创建一个数字编辑框。在项目"UIDemo"中,首先用"<!-- 代码段 -->"注释掉Button标签代码,然后添加如下代码:

```
<EditText
    android:layout_width="100dp"
    android:layout_height="wrap_content"
    android:inputType="number"
/>
```

运行程序,可看到结果如图2.13所示,当用户点击编辑框时,弹出小键盘中只有数字键盘允许选择输入。

图2.13 数字编辑框

【例2.6】创建一个带提示信息、允许用户输入最长16个字符的密码框。在项目"UIDemo"中,首先用"<!-- 代码段 -->"注释掉上例中的<EditTex>标签部分,然后添加如下代码:

```
<EditText
    android:layout_width="match_parent"
    android:layout_height="wrap_content"
    android:hint="请输入密码,最长16位"
    android:inputType="textPassword"
    android:maxLength="16"/>
```

运行程序,可看到结果如图2.14所示,默认状态下文本框中以浅色字显示"请输入密码,最长16位",用户输入后,字符以密码格式显示出来。

图2.14 密码框

### 2.2.6 复选框(CheckBox)

复选框提供"选中"和"未选中"两种状态供用户选择,它继承了按钮的多数外观属性,并提供了选择状态的属性,其常用属性如表2.6所示。

表2.6 复选框常用属性列表

| 属性声明 | 功能描述 | 属性值 |
| --- | --- | --- |
| Android:checked | 设置是否被选中 | "true"选择<br>"false"未选中 |

[例2.7] 创建一个复选框。在项目"UIDemo"中,首先用"<!-- 代码段 -->"注释掉其他组件标签的代码段,然后添加如下代码:

```
<CheckBox
    android:layout_width="match_parent"
    android:layout_height="wrap_content"
    android:checked="true"
    android:text="中国"/>
```

运行程序,可看到结果如图2.15所示,当前复选框为选中状态,用户可以通过点击来改变状态。

图2.15 复选框

### 2.2.7 单选按钮(RadioButton)

单选按钮也提供"选中"和"未选中"两种状态供用户选择,与复选框不同的是,它可以与RadioGroup组合使用,使得一组单选按钮中只能有一个被选中。RadioGroup用来管理单选按钮组,RadioButton用来进行选择,它们的常用属性如表2.7所示。

表 2.7 RadioButton 和 RadioGroup 常用属性列表

| 元素 | 属性声明 | 功能描述 | 属性值 |
| --- | --- | --- | --- |
| RadioGroup | android:orientation | 设置内部多个单选按钮排列方式,水平或者垂直排列 | "horizontal"水平布局 "vertical"垂直布局 |
| RadioButton | android:checked | 设置是否被选中 | "true"选中,"false"未选中 |

[例 2.8] 创建一个用来进行"男""女"性别选择的单选框。在项目"UIDemo"中,首先用"<!-- 代码段 -->"注释掉其他组件标签的代码段,然后添加如下代码:

```
<RadioGroup
    android:layout_width="wrap_content"
    android:layout_height="wrap_content"
    android:orientation="horizontal">
    <RadioButton
        android:layout_width="wrap_content"
        android:layout_height="wrap_content"
        android:text="男"
        android:textSize="20sp"/>
    <RadioButton
        android:layout_width="wrap_content"
        android:layout_height="wrap_content"
        android:text="女"
        android:textSize="20sp"/>
</RadioGroup>
```

运行程序,可看到结果如图 2.16 所示,此时,只有"男"或者"女"之一能被选中,选中后颜色和样式发生变化。

图 2.16 单选按钮

### 2.2.8 状态切换按钮(ToggleButton)

状态切换按钮是一个用来设置"开"和"关"两种状态的特殊按钮,并可以根据这两种状态显示不同文本标题或者图片效果,其常用属性如表 2.8 所示。

## 表2.8 状态切换按钮常用属性列表

| 属性声明 | 功能描述 | 属性值 |
| --- | --- | --- |
| android:textOff | 设置关闭状态时显示的文字 | 字符串,如"关闭""停止" |
| android:textOn | 设置打开状态时显示的文字 | 字符串,如"打开""启动" |
| android:disabledAlpha | 设置按钮禁用时的透明度 | [0,1],默认值为0.5 |

【例2.9】创建一个简单的状态切换按钮。在项目"UIDemo"中,首先用"<!-- 代码段 -->"注释掉其他组件标签的代码段,然后添加如下代码:

```
<ToggleButton
    android:layout_width="wrap_content"
    android:layout_height="wrap_content"
    android:textOff="关闭"
    android:textOn="打开"
    android:textSize="20sp"
    android:disabledAlpha="0.5"/>
```

运行程序,可看到结果如图2.17所示,点击按钮,状态会在"关闭"和"打开"之间切换。

图2.17 简单状态切换按钮

【例2.10】创建一个具有左右切换效果的状态开关按钮。在项目"UIDemo"中,首先用"<!-- 代码段 -->"注释掉其他组件标签的代码段,然后进行以下步骤:

(1) 设计两张按钮状态效果图"ic_off.png"和"ic_on.png",如图2.18所示,分别代表打开和关闭状态,把图片导入项目的drawable目录中。

图2.18 状态效果图

(2) 创建<selector>元素文件"tb_selector.xml",针对state_checked属性的不同状态定义<item>。

```
<?xml version="1.0" encoding="utf-8"?>
<selector xmlns:android="http://schemas.android.com/apk/res/android">
    <item android:drawable="@drawable/ic_off" android:state_checked="false"/>
    <item android:drawable="@drawable/ic_on" android:state_checked="true"/>
</selector>
```

（3）在activity_main.xml布局文件中注释掉RelativeLayout标记中的代码片段，在其中添加如下代码：

```
<ToggleButton
    android:layout_width="90dp"
    android:layout_height="20dp"
    android:textOff="@null"
    android:textOn="@null"
    android:background="@drawable/tb_selector"/>
```

运行程序，可看到结果如图2.19所示，点击按钮，状态切换。

图2.19 左右切换效果的状态开关按钮

Android还提供了Switch（状态开关按钮）用来定义更个性化的状态开关按钮，Switch除了具有和ToggleButton相同的属性外，还提供了android:textStyle属性设置文本风格、android:thumb属性指定自定义Drawable绘制开关按钮、android:track属性指定自定义Drawable绘制开关轨道、android:typeface属性设置字体风格等，读者可自行测试练习。

### 2.2.9 下拉列表（Spinner）

下拉列表以下拉项的方式为用户提供选择项。下拉列表常用的属性如表2.9所示。

表2.9 下拉列表常用属性列表

| 属性声明 | 功能描述 | 属性值 |
| --- | --- | --- |
| android:entries | 绑定XML文件中的数据源 | 数据源引用，如"@array/cost"，cost为已在XML资源文件中创建好的资源数组的名称 |
| android:spinnerMode | 显示模式 | "dropdown"直接下拉模式，"dialog"对话框效果模式 |
| android:dropDownWidth | 下拉框宽度 | 像素值，dp为单位，如"30dp" |

【例2.11】创建一个下拉列表。在项目"UIDemo"中，首先用"<!-- 代码段 -->"注释掉其他组件标签的代码段，然后进行以下步骤：

（1）在项目"UIDemo"的res/values目录上点击右键，选择"New"→"XML"→"Values XML File"，在弹出界面中输入文件名"arrays"，创建arrays.xml文件，如图2.20所示。

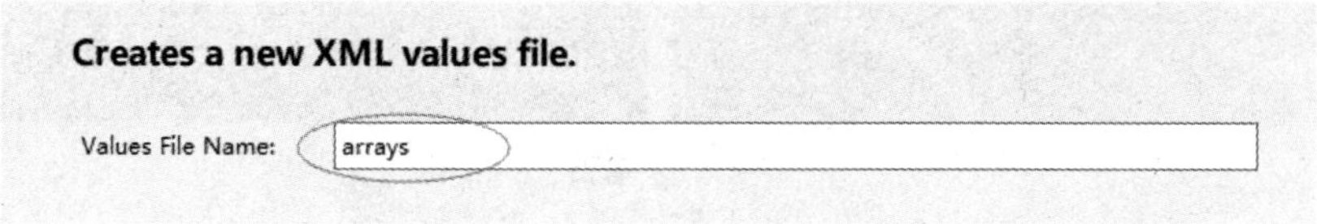

图2.20 创建arrays资源文件

文件初始时结构如下：

```xml
<?xml version="1.0" encoding="utf-8"?>
<resources>
</resources>
```

（2）在arrays.xml文件的<resources>标签中加入选项列表，代码如下：

```xml
<resources>
    <string-array name="cost">
        <item>餐费</item>
        <item>交通费</item>
        <item>住宿费</item>
    </string-array>
</resources>
```

<string-array>标签为Android提供了一种简单的资源数组定义方式，可以很直接地在XML文件中定义数据项相对固定的资源数据。其name属性作为它在布局文件或者Java代码中的引用标识。

（3）在activity_main.xml布局文件中添加如下代码：

```xml
<Spinner
    android:layout_width="match_parent"
    android:layout_height="wrap_content"
    android:spinnerMode="dropdown"
    android:dropDownWidth="100dp"
    android:entries="@array/cost"/>
```

@array/name用来在布局文件中引用自定义的<string-array>资源数据，如本例中的"@array/cost"。

运行程序，可以看到如图2.21所示效果，点击下拉箭头，可以显示出列表项。

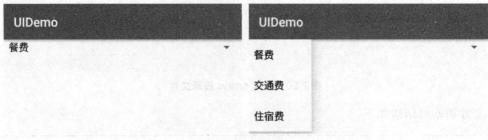

图 2.21 下拉列表

### 2.2.10 图片视图（ImageView）

图片视图主要用来显示图片和图片类的对象。图片视图常用的属性如表 2.10 所示。

表 2.10 图片视图常用属性列表

| 属性声明 | 功能描述 | 属性值 |
| --- | --- | --- |
| android:src | 设置显示的图片 | 图片数据源，如"@drawable/ic_launcher"，ic_launcher 为图片对象的 ID |
| android:scaleType | 调整图片缩放、位置等以满足 ImageView 显示的需要 | "matrix"：图片的宽度/高度（自左上角）大于 ImageView，剪裁掉超出部分，否则，不做处理<br>"fitXY"：图片填充满整个 ImageView<br>"fitStart"：图片按比例缩放至 ImageView 的宽度和高度中小的那个值，然后居上或者居左显示（视按宽或者高缩放而定）<br>"fitCenter"：与"fitStart"相似，只是显示方式为居中<br>"fitEnd"：与"fitStart"相似，只是显示方式为居下或者居右（视按宽或者高缩放而定）<br>"center"：图片居中显示，如果超出了 ImageView 的大小，则剪裁掉四周超出的部分<br>"centerCrop"：按比例缩放填充满 ImageView，并将多余的宽或高出部分剪裁掉<br>"centerInside"按比例缩放图片，在 ImageView 中完整显示，空出部分填充背景色 |

[例 2.12] 通过图片视图显示一张图片。在项目"UIDemo"中，首先用"<!-- 代码段 -->"注释掉其他组件标签的代码段，然后进行以下步骤：

（1）把一张图片"flower.png"复制到 res/drawable 目录下。

（2）在 activity_main.xml 布局文件中添加如下代码：

```
<ImageView
    android:layout_width="wrap_content"
    android:layout_height="wrap_content"
```

android:src="@drawable/flower"
android:scaleType="center"/>

运行程序,可看到结果如图2.22所示。

图2.22　图片视图

### 2.2.11　图片按钮(ImageButton)

图片按钮继承自图片视图,它以图片或图片Drawable对象来设置显示外观,与按钮不同的是,图片按钮通过"android:src"属性设置的图片是前景图,区别于"android:background"属性设置的背景图,同时,图片按钮没有"android:text"属性。图片按钮常用的属性如表2.11所示。

表2.11　图片按钮常用属性列表

| 属性声明 | 功能描述 | 属性值 |
| --- | --- | --- |
| android:src | 设置图片按钮显示的图片或者图片对象 | 图片数据源,如"@drawable/ic_add",ic_add图片名称或者图片Drawable对象的ID属性 |
| android:scaleType | 调整图片的缩放、位置等,以满足ImageView显示的需要 | 参见表2.10,图片视图常用属性列表 |

[例2.13] 创建一个图片按钮。在项目"UIDemo"中,首先用"<!-- 代码段 -->"注释掉其他组件标签的代码段,然后进行以下步骤:

(1)把一张图片"ic_add.png"复制到res/drawable目录下。

(2)在activity_main.xml布局文件中添加如下代码:

```
<ImageButton
    android:layout_width="wrap_content"
    android:layout_height="wrap_content"
    android:src="@drawable/ic_add"
    android:background="#317EF3"/>
```

运行程序,可看到效果如图2.23所示。

### 2.2.12 其他UI组件

Android还提供了其他功能丰富的UI组件,如日期选择器(DatePicker)、时间选择器(TimePicker)、进度条(ProgressBar)、拖动条(SeekBar)等,掌握这些组件的基本用法,可以实现丰富的界面效果,例如:

(1)日期选择器,如图2.24所示,代码如下:

图2.23 图片按钮　　　　　　　　图2.24 日期选择器

```
<DatePicker
    android:layout_width="wrap_content"
    android:layout_height="150dp"
    android:calendarViewShown="false"
    android:datePickerMode="spinner"
/>
```

(2)时间选择器,如图2.25所示,代码如下:

```
<TimePicker
    android:layout_width="wrap_content"
    android:layout_height="wrap_content"/>
```

图 2.25　时间选择器

（3）进度条，如图 2.26 所示，代码如下：

```
<ProgressBar
    android:layout_width="match_parent"
    android:layout_height="wrap_content"/>
```

（4）拖动条，如图 2.27 所示，代码如下：

```
<SeekBar
    android:layout_width="match_parent"
    android:layout_height="wrap_content"/>
```

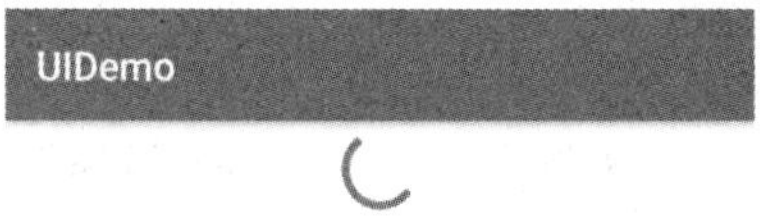

图 2.26　进度条　　　　　　　　　图 2.27　拖动条

## 2.3　布局

布局是 Android 提供的 ViewGroup 类的子类，作为容器，UI 组件可以按照一定规则放置在布局中，布局本身也可以作为组件嵌套使用，以完成多种界面设计的需要。Android 提供的主要布局类型包括相对布局（RelativeLayout）、线性布局（LinearLayout）、框架布局（FrameLayout）、表格布局（TableLayout）、网格布局（GridLayout）、绝对布局（AbsoluteLayout）、约束布局（ConstraintLayout）。

### 2.3.1 相对布局(RelativeLayout)

相对布局(RelativeLayout)通过提供组件相对位置的描述信息来对组件进行定位,即需要一个参照物来对组件进行定位。参照物可以是布局(该组件的父容器),也可以是布局中的其他组件(该组件的兄弟组件)。例如:一个文本标签被放置在父容器的右对齐位置,一个按钮被放置在一个文本标签的下方。

相对布局通过相对关系来进行布局,在界面设计中可以减少布局嵌套层次、提升灵活性,在外层布局中可以使用这种布局方式。

相对布局的使用方法如下:

(1)先为作为参照物的组件(或者容器)声明id属性。如为一个文本标签定义属性:android:id="@+id/tv_name"。

(2)在需要参照定位的组件属性声明中,指定具体位置关系进行定位。如一个按钮被声明放置在id为"tv_name"的文本标签下方:android:layout_below="@id/tv_name"。"@id/name"表示引用id为name的组件。

注意,作为第一个参照物的组件,其定位一般需要参照父容器,例如定位在父容器的左上角:

android:layout_alignParentTop="true"

android:layout_alignParentLeft="true"

(3)通过设置组件的外边距、内边距调整间隔,通过设置组件的对齐方式调整定位。如:

android:layout_marginTop="100dp"

android:layout_alignRight="@id/tv_name"

Android提供了一些定位属性供布局中的组件使用,如表2.12所示。

表2.12 相对布局常用属性列表

| 属性声明 | 功能描述 | 取值范围 |
| --- | --- | --- |
| android:layout_alignParentLeft | 当前组件是否与其父容器左对齐 | "true""false" |
| android:layout_alignParentTop | 当前组件是否与其父容器顶部对齐 | "true""false" |
| android:layout_alignParentRight | 当前组件是否与其父容器右对齐 | "true""false" |
| android:layout_alignParentBottom | 当前组件是否与其父容器底部对齐 | "true""false" |
| android:layout_centerHorizontal | 当前组件是否在其父容器中水平居中 | "true""false" |
| android:layout_centerVertical | 当前组件是否在其父容器中垂直居中 | "true""false" |
| android:layout_centeInparent | 当前组件是否在其父容器中居中 | "true""false" |
| android:layout_toRightOf | 当前组件放到指定id组件的右边 | 组件id |
| android:layout_toLeftOf | 当前组件放到指定id组件的左边 | 组件id |
| android:layout_above | 当前组件放到指定id组件的上边 | 组件id |
| android:layout_below | 当前组件放到指定id组件的下边 | 组件id |
| android:layout_alignBaseline | 当前组件与指定id组件基准线对齐 | 组件id |
| android:layout_alignLeft | 当前组件与指定id组件左对齐 | 组件id |
| android:layout_alignRight | 当前组件与指定id组件右对齐 | 组件id |
| android:layout_alignTop | 当前组件与指定id组件顶部对齐 | 组件id |
| android:layout_alignBottom | 当前组件与指定id组件底部对齐 | 组件id |

[例2.14] 相对布局简单示例。在项目"UIDemo"中,首先自定义一个基于相对布局的 XML 布局文件,然后设置该文件为当前显示视图。

(1) 在 res/layout 目录上点击右键,选择"New"→"Layout resource file",填写"File name"和"Root element",如图 2.28 所示。注意,"File name"由小写字母、下划线和数字组成,"Root element"中的"RelativeLayout"区分大小写。

图 2.28 创建相对布局文件

(2) 打开 res/layout 目录中新生成的 ly_relativelayout.xml 文件,点击左下角的"Text"视图模式,此时可以看到文档的布局方式为 RelativeLayout。

```xml
<?xml version="1.0" encoding="utf-8"?>
<RelativeLayout xmlns:android="http://schemas.android.com/apk/res/android"
    android:layout_width="match_parent" android:layout_height="match_parent">
</RelativeLayout>
```

(3) 在 RelativeLayout 布局下添加如下代码:

```xml
<?xml version="1.0" encoding="utf-8"?>
<RelativeLayout xmlns:android="http://schemas.android.com/apk/res/android"
    android:layout_width="match_parent" android:layout_height="match_parent">
    <EditText
        android:id="@+id/edt_input"
        android:layout_width="200dp"
        android:layout_height="40dp"
        android:inputType="text"
        android:layout_alignParentTop="true"
        android:layout_alignParentLeft="true"/>
    <Button
        android:id="@+id/btn_select"
        android:layout_width="wrap_content"
        android:layout_height="40dp"
        android:text="查询"
        android:layout_toRightOf="@id/edt_input"
        android:layout_alignBottom="@id/edt_input"
        />
</RelativeLayout>
```

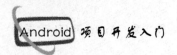

（4）打开java/cc.turbosnail.uidemo/MainActivity，修改MainActivity.java的代码如下：

```
protected void onCreate(Bundle savedInstanceState) {
    super.onCreate(savedInstanceState);
    //setContentView(R.layout.activity_main);
    setContentView(R.layout.ly_relativelayout);//设置ly_relativelayout为当前显示视图
}
```

运行程序，效果如图2.29所示。

图2.29 相对布局简单示例

[项目案例1] 设计并实现个随身账本项目的用户注册界面

实现步骤：

（1）在Android Studio菜单栏中点击"File"→"New"→"New Project"，创建新项目，项目名称为"AccountBook"，如图2.30所示。

图2.30 创建AccountBook案例项目

（2）导入资源文件。把"ic_icon.png""ic_message.png""ic_password.png""ic_username.png"等4个图标文件复制到res/drawable目录下，这些图标将作为界面元素美化界面使用。

（3）新建布局文件。在res/layout目录上点击右键，选择"New"→"Layout resource file"，填写"File name"为"activity_register"，"Root element"为"RelativeLayout"，如图2.31所示。

图2.31 新建注册界面布局文件

（4）编写布局代码。打开res/layout/activity_register.xml文件，编写如下布局代码：

```
<?xml version="1.0" encoding="utf-8"?>
<RelativeLayout xmlns:android="http://schemas.android.com/apk/res/android"
    android:layout_width="match_parent" android:layout_height="match_parent">
    <ImageView
        android:id="@+id/iv_icon_register"
        android:layout_width="wrap_content"
        android:layout_height="wrap_content"
        android:layout_centerHorizontal="true"
        android:layout_marginTop="20dp"
        android:src="@drawable/ic_icon"/>
    <TextView
        android:id="@+id/tv_username_register"
        android:layout_width="wrap_content"
        android:layout_height="wrap_content"
        android:layout_below="@id/iv_icon_register"
        android:layout_marginTop="60dp"
        android:layout_marginLeft="60dp"
        android:text="用户名:"
        android:textSize="16sp"
        android:textColor="#000000"/>
    <ImageView
        android:id="@+id/iv_username_register"
        android:layout_width="20dp"
        android:layout_height="20dp"
        android:src="@drawable/ic_username"
        android:layout_below="@id/iv_icon_register"
        android:layout_alignTop="@id/tv_username_register"
        android:layout_toRightOf="@id/tv_username_register"/>
    <EditText
        android:id="@+id/edt_username_register"
        android:layout_width="180dp"
```

```xml
        android:layout_height="wrap_content"
        android:inputType="text"
        android:hint="请输入用户名"
        android:layout_below="@id/iv_icon_register"
        android:layout_marginTop="50dp"
        android:layout_toRightOf="@id/iv_username_register"/>
    <TextView
        android:id="@+id/tv_password_register"
        android:layout_width="wrap_content"
        android:layout_height="wrap_content"
        android:layout_below="@id/iv_icon_register"
        android:layout_marginTop="100dp"
        android:layout_alignRight="@id/tv_username_register"
        android:text="密码:"
        android:textSize="16sp"
        android:textColor="#000000"/>
    <ImageView
        android:id="@+id/iv_password_register"
        android:layout_width="20dp"
        android:layout_height="20dp"
        android:src="@drawable/ic_password"
        android:layout_below="@id/iv_icon_register"
        android:layout_alignTop="@id/tv_password_register"
        android:layout_toRightOf="@id/tv_password_register"/>
    <EditText
        android:id="@+id/edt_password_register"
        android:layout_width="180dp"
        android:layout_height="wrap_content"
        android:inputType="numberPassword"
        android:hint="请输入密码"
        android:layout_below="@id/iv_icon_register"
        android:layout_marginTop="90dp"
        android:layout_toRightOf="@id/iv_password_register"/>
    <TextView
        android:id="@+id/tv_repeat_register"
        android:layout_width="wrap_content"
        android:layout_height="wrap_content"
        android:layout_below="@id/iv_icon_register"
        android:layout_marginTop="140dp"
```

```
            android:layout_alignRight="@id/tv_username_register"
            android:text="确认密码:"
            android:textSize="16sp"
            android:textColor="#000000"/>
        <ImageView
            android:id="@+id/iv_repeat_register"
            android:layout_width="20dp"
            android:layout_height="20dp"
            android:src="@drawable/ic_password"
            android:layout_below="@id/iv_icon_register"
            android:layout_alignTop="@id/tv_repeat_register"
            android:layout_toRightOf="@id/tv_repeat_register"/>
        <EditText
            android:id="@+id/edt_repeat_register"
            android:layout_width="180dp"
            android:layout_height="wrap_content"
            android:inputType="numberPassword"
            android:hint="请确认密码"
            android:layout_below="@id/iv_icon_register"
            android:layout_marginTop="130dp"
            android:layout_toRightOf="@id/iv_repeat_register"/>
        <Button
            android:id="@+id/btn_register"
            android:layout_width="200dp"
            android:layout_height="40dp"
            android:text="注册"
            android:textSize="18sp"
            android:textColor="#FFFFFF"
            android:background="@color/colorPrimary"
            android:layout_below="@id/iv_icon_register"
            android:layout_centerHorizontal="true"
            android:layout_marginTop="220dp"/>
        <TextView
            android:id="@+id/tv_login_register"
            android:layout_width="wrap_content"
            android:layout_height="wrap_content"
            android:text="已有账号,前去登录~~"
            android:textSize="14sp"
            android:textColor="@color/colorPrimary"
```

android:layout_below="@id/btn_register"

android:layout_centerHorizontal="true"

android:layout_marginTop="40dp"/>

</RelativeLayout>

（5）打开java/cc.turbosnail.accountbook/MainActivity，修改代码如下，并运行程序：

```
protected void onCreate(Bundle savedInstanceState) {
    super.onCreate(savedInstanceState);
    //setContentView(R.layout.activity_main);
    setContentView(R.layout.activity_register);//设置activity_register为当前显示视图
}
```

程序运行效果如图2.32所示。

图2.32　注册界面运行效果图

### 2.3.2　线性布局（LinearLayout）

线性布局对组件按照添加的先后顺序依次排列的方式进行定位，所有组件被排列在一行或者一列上，超出布局范围的部分不会自动换行或者换列，即超出后不会显示出来。线性布局提供android:orientation属性来决定其内部组件是水平一行显示，还是垂直一列显示。线性布局在需要顺次排列组件时使用，其常用属性如表2.13所示。

表 2.13 线性布局常用属性列表

| 属性声明 | 功能描述 | 取值范围 |
| --- | --- | --- |
| android:orientation | 在布局中声明组件的排列方式 | "horizontal",默认值,水平排列,所有组件按一行排列<br>"vertical",垂直排列,所有组件按一列排列 |
| android:layout_gravity | 当前组件在线性布局中的对齐方式 | "center",居中对齐<br>"clip_horizontal",水平方向对齐;"clip_vertical",垂直方向对齐<br>在水平排列中可使用"bottom""top""center_vertical",在垂直排列中可使用"right""left""center_horizontal"<br>多个值组合可以用"\|"分隔,如"top\|clip_vertical" |

【例2.15】线性布局简单示例。在项目"UIDemo"中,首先自定义一个基于线性布局的 XML 布局文件,然后设置该文件为当前显示视图。

(1) 在 res/layout 目录上点击右键,选择"New"→"Layout Resource file",填写"File name"为"ly_linearlayout","Root element"为"LinearLayout",如图 2.33 所示。

图 2.33 创建线性布局文件

(2) 编写布局代码。打开 res/layout/ly_linearlayout.xml 文件,切换到"text"视图模式,编写如下布局代码:

```
<?xml version="1.0" encoding="utf-8"?>
<LinearLayout xmlns:android="http://schemas.android.com/apk/res/android"
    android:orientation="horizontal" android:layout_width="match_parent"
    android:layout_height="match_parent">
    <Button
        android:layout_width="110dp"
        android:layout_height="wrap_content"
        android:text="按钮 1"/>
    <Button
        android:layout_width="110dp"
        android:layout_height="wrap_content"
        android:text="按钮 2"/>
    <Button
        android:layout_width="110dp"
```

android:layout_height="wrap_content"

android:text="按钮3"/>

&lt;Button

android:layout_width="110dp"

android:layout_height="wrap_content"

android:text="按钮4"/>

&lt;/LinearLayout>

(3) 打开java/cc.turbosnail.uidemo/MainActivity,修改代码如下：

protected void onCreate(Bundle savedInstanceState) {

super.onCreate(savedInstanceState);

//setContentView(R.layout.activity_main);

setContentView(R.layout.ly_linearlayout);//设置ly_linearlayout为当前显示视图

}

程序运行效果如图2.34所示,第四个按钮超出屏幕部分被遮挡住。

图2.34　水平线性布局示例

如果将LinearLayout中android：orientation属性的值改为"vertical",即android：orientation="vertical",便会产生如图2.35所示效果。

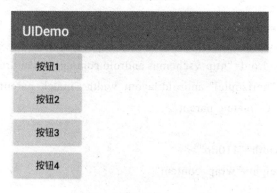

图2.35　垂直线性布局示例

### 2.3.3 帧布局(FrameLayout)

帧布局把所有添加其中的组件按照先后顺序层叠起来显示,组件的默认位置都是在布局的左上角,第一个添加的组件在最底层。如果添加的组件都大小相同,那么同一时刻就只能看到最上面的组件。在帧布局中,组件可以和在线性布局中一样,通过android：

layout_gravity属性调整对齐方式。

[例2.16] 帧布局简单示例。在项目"UIDemo"中,首先自定义一个基于帧布局的XML布局文件,然后设置该文件为当前显示视图。

(1) 在res/layout目录上点击右键,选择"New"→"Layout Resource file",填写"File name"为"ly_framelayout","Root element"为"FrameLayout",如图2.36所示。

图2.36 创建帧布局文件

(2) 编写布局代码。打开res/layout/ly_framelayout.xml文件,切换到"text"视图模式,编写如下布局代码。其中,"ic_message.png"图标文件需要先复制到drawable目录下。

```xml
<?xml version="1.0" encoding="utf-8"?>
<FrameLayout xmlns:android="http://schemas.android.com/apk/res/android"
    android:layout_width="match_parent" android:layout_height="match_parent">
    <ImageView
        android:layout_width="wrap_content"
        android:layout_height="wrap_content"
        android:src="@drawable/ic_message"/>
    <TextView
        android:layout_width="wrap_content"
        android:layout_height="wrap_content"
        android:layout_marginLeft="90dp"
        android:layout_marginTop="20dp"
        android:text="99+"
        android:textColor="@color/colorAccent"
        android:textSize="18sp"/>
</FrameLayout>
```

(3) 打开java/cc.turbosnail.uidemo/MainActivity,修改代码如下:

```java
protected void onCreate(Bundle savedInstanceState) {
    super.onCreate(savedInstanceState);
    //setContentView(R.layout.activity_main);
    setContentView(R.layout.ly_framelayout);//设置ly_framelayout为当前显示视图
}
```

运行程序,效果如图2.37所示,通过调整文本标签的外边距,使得它与图片视图的右上角重合,模拟出一个角标的效果。

图2.37　帧布局示例

### 2.3.4　表格布局(TableLayout)

表格布局采用行、列的形式来放置组件,在布局中添加 TableRow 标签表示行,一个 TableRow 标签表示一行,每一行中可以多个添加 UI 组件,一个 UI 组件占据一列,列宽由当前列中最宽的 UI 组件确定。表格布局常用的属性如表2.14所示。

表2.14　表格布局常用属性

| 属性声明 | 功能描述 | 取值范围 |
| --- | --- | --- |
| android:collapseColumns | 设置需要被隐藏不显示的列 | 列编号,从"0"开始,多个列号用逗号隔开,"*"表示所有列,如 android:collapseColumns="0,3" 把第一列和第四列隐藏不显示 |
| android:shrinkColumns | 设置可自动收缩的列,以保证当前行能在容器行中全部显示 | android:shrinkColumns="1" |
| android:stretchColumns | 设置可自动拉伸的列,以保证当前行占满容器整行 | 当一行中组件超出行宽后,第二列自动收缩变小 |

表格布局中添加的组件可以通过设置自身的属性来指定其在表格中占据的位置,组件常用属性如表2.15所示。

表2.15　表格布局中的组件常用属性

| 属性声明 | 功能描述 | 示例 |
| --- | --- | --- |
| android:layout_span | 设置该组件占据的列数,默认值是"1" | &lt;TextView android:text="总计金额" |
| android:layout_column | 设置该组件放置在第几列,列编号从"0"开始 | android:layout_column="1" android:layout_span="2"/&gt; |

[例2.17] 表格布局简单示例。在项目"UIDemo"中,首先自定义一个基于表格布局的 XML 布局文件,然后设置该文件为当前显示视图。

(1) 在 res/layout 目录上点击右键,选择"New"→"Layout Resource file",填写"File name"为"ly_tablelayout","Root element"为"TableLayout",如图2.38所示。

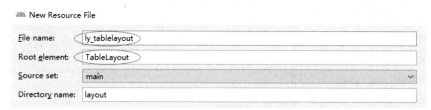

图2.38 创建表格布局文件

（2）编写布局代码。打开res/layout/ly_tablelayout.xml文件，切换到"text"视图模式，编写如下布局代码：

```
<?xml version="1.0" encoding="utf-8"?>
<TableLayout xmlns:android="http://schemas.android.com/apk/res/android"
    android:layout_width="match_parent" android:layout_height="match_parent"
    android:stretchColumns="*"
    android:background="#606060">
    <TableRow
        android:layout_width="match_parent"
        android:layout_height="match_parent">
        <TextView
            android:layout_width="wrap_content"
            android:layout_height="wrap_content"
            android:text="序号"
            android:textColor="#000"
            android:textSize="20sp"
            android:gravity="center"
            android:background="#FFF"
            android:layout_margin="0.5dp"/>
        <TextView
            android:layout_width="wrap_content"
            android:layout_height="wrap_content"
            android:text="类别"
            android:textColor="#000"
            android:textSize="20sp"
            android:gravity="center"
            android:background="#FFF"
            android:layout_margin="0.5dp"/>
        <TextView
            android:layout_width="wrap_content"
            android:layout_height="wrap_content"
            android:text="金额"
```

```xml
            android:textColor="#000"
            android:textSize="20sp"
            android:gravity="center"
            android:background="#FFF"
            android:layout_margin="0.5dp"/>
    </TableRow>
    <TableRow
        android:layout_width="match_parent"
        android:layout_height="match_parent">
        <TextView
            android:layout_width="wrap_content"
            android:layout_height="wrap_content"
            android:text="1"
            android:textColor="#000"
            android:textSize="18sp"
            android:gravity="center"
            android:background="#FFF"
            android:layout_margin="0.5dp"/>
        <TextView
            android:layout_width="wrap_content"
            android:layout_height="wrap_content"
            android:text="早餐"
            android:textColor="#000"
            android:textSize="18sp"
            android:gravity="center"
            android:background="#FFF"
            android:layout_margin="0.5dp"/>
        <TextView
            android:layout_width="wrap_content"
            android:layout_height="wrap_content"
            android:text="20"
            android:textColor="#000"
            android:textSize="18sp"
            android:gravity="center"
            android:background="#FFF"
            android:layout_margin="0.5dp"/>
    </TableRow>
    <TableRow
        android:layout_width="match_parent"
```

```xml
        android:layout_height="match_parent">
        <TextView
            android:layout_width="wrap_content"
            android:layout_height="wrap_content"
            android:text="2"
            android:textColor="#000"
            android:textSize="18sp"
            android:gravity="center"
            android:background="#FFF"
            android:layout_margin="0.5dp"/>
        <TextView
            android:layout_width="wrap_content"
            android:layout_height="wrap_content"
            android:text="交通"
            android:textColor="#000"
            android:textSize="18sp"
            android:gravity="center"
            android:background="#FFF"
            android:layout_margin="0.5dp"/>
        <TextView
            android:layout_width="wrap_content"
            android:layout_height="wrap_content"
            android:text="30"
            android:textColor="#000"
            android:textSize="18sp"
            android:gravity="center"
            android:background="#FFF"
            android:layout_margin="0.5dp"/>
    </TableRow>
    <TableRow
        android:layout_width="match_parent"
        android:layout_height="match_parent">
        <TextView
            android:layout_width="wrap_content"
            android:layout_height="wrap_content"
            android:layout_span="2"
            android:text="合计"
            android:textColor="#000"
            android:textSize="20sp"
```

```xml
            android:gravity="center"
            android:background="#FFF"
            android:layout_margin="0.5dp"/>
        <TextView
            android:layout_width="wrap_content"
            android:layout_height="wrap_content"
            android:text="50"
            android:textColor="#000"
            android:textSize="20sp"
            android:gravity="center"
            android:background="#FFF"
            android:layout_margin="0.5dp"/>
    </TableRow>
</TableLayout>
```

(3)打开java/cc.turbosnail.uidemo/MainActivity,修改代码如下:

```java
protected void onCreate(Bundle savedInstanceState) {
    super.onCreate(SavedInstanceState);
    //setContentView(R.layout.activity_main);
    setContentView(R.layout.ly_tablelayout);//设置ly_tablelayout为当前显示视图
}
```

运行程序,效果如图2.39所示。

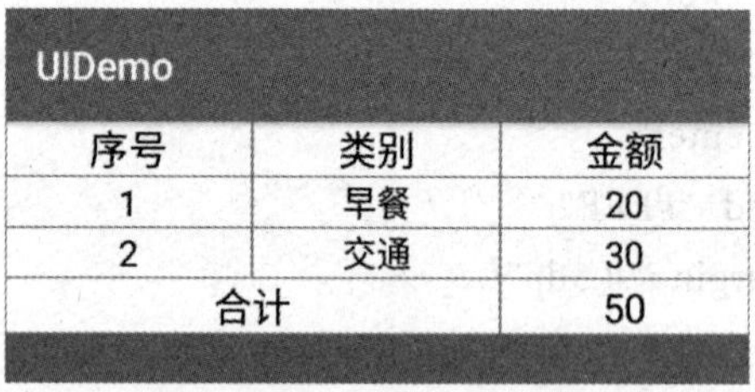

| UIDemo | | |
|---|---|---|
| 序号 | 类别 | 金额 |
| 1 | 早餐 | 20 |
| 2 | 交通 | 30 |
| 合计 | | 50 |

图2.39 表格布局示例

注意:表格布局默认是不带边框的,本例中通过为表格布局和TextView设置不同的背景颜色,并通过为TextView设置0.5 dp的外边距(android:layout_margin)的方式实现了边框效果。另外,本段代码中出现大量重复的属性设置,可以通过把这些属性自定义成一个样式(style)文件来引用,使代码更简洁。关于样式的使用,请读者参见本书第4章内容。

### 2.3.5 网格布局(GridLayout)

网格布局通过指定行数和列数将布局划分成多个单元格,每个单元格可以放置一个UI组件。

与表格布局相比,网格布局更加灵活。表格布局的列数与放置的内容关联,是由组件的个数决定的,而网格布局可以根据需要划分好行和列,并设置行和列的各种属性,再添加组件。网格布局还可以通过指定行号和列号把组件直接放置到对应单元格中,一个组件也可以横跨多个行或者列显示。

在放置组件时,与线性布局一样,网格布局也分为水平和垂直两种方式,默认是水平方式,即组件从左到右依次排列。网格布局常用的属性如表2.16所示。

表2.16 网格布局常用属性

| 属性声明 | 功能描述 | 取值范围 |
| --- | --- | --- |
| android:columnCount | 设置总的列数 | 整数值,如"3" |
| android:rowCount | 设置总的行数 | 整数值,如"5" |
| android:orientation | 子元素的布局方向 | "horizontal":水平布局<br>"vertical":竖直布局 |

网格布局中的组件可以通过添加属性来指定其占据的位置,常用属性如表2.17所示。

表2.17 网格布局中的组件常用属性

| 属性声明 | 功能描述 | 取值范围 |
| --- | --- | --- |
| android:layout_columnSpan | 设置该组件占用的列数 | 整数值,如当前组件占2列显示:<br>android:layout_columnSpan="2" |
| android:layout_rowSpan | 设置该组件占用的行数 | 整数值,如当前组件占2行显示:<br>android:layout_rowSpan="2" |
| android:layout_column | 设置该组件显示在第几列,列编号从"0"开始 | 整数值,如把组件放在第2列第3行:<br>android:layout_column="1" |
| android:layout_row | 设置该组件显示在第几行,行编号从"0"开始 | android:layout_row="2" |
| android:layout_columnWeight<br>android:layout_rowWeight | 设置该组件的列/行权重 | 整数值,如GridLayout有3列,添加3个组件,每个组件都设置android:layout_columnWeight="1",则3个组件按相同权重平分行空间 |

[例2.18] 网格布局简单示例:计算器界面设计。在项目"UIDemo"中,自定义一个基于网格布局的XML布局文件,然后设置该文件为当前显示视图。

(1) 在res/layout目录上点击右键,选择"New"→"Layout Resource file",填写File name为"ly_gridlayout",Root element为"GridLayout",如图2.40所示。

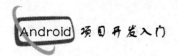

图2.40 创建网格布局文件

（2）编写布局代码。打开res/layout/ly_gridlayout.xml文件，切换到"text"视图模式，编写如下布局代码：

```
<?xml version="1.0" encoding="utf-8"?>
<GridLayout xmlns:android="http://schemas.android.com/apk/res/android"
    android:layout_width="match_parent"
    android:layout_height="match_parent"
    android:columnCount="4"
    android:rowCount="5">
    <EditText
        android:layout_width="match_parent"
        android:layout_height="100dp"
        android:layout_columnSpan="4"/>
    <Button
        android:layout_columnWeight="1"
        android:layout_rowWeight="1"
        android:text="清空"/>
    <Button
        android:layout_columnWeight="1"
        android:layout_rowWeight="1"
        android:text="退格"/>
    <Button
        android:layout_columnWeight="1"
        android:layout_rowWeight="1"
        android:text="/"/>
    <Button
        android:layout_columnWeight="1"
        android:layout_rowWeight="1"
        android:text="*"/>
    <Button
        android:layout_columnWeight="1"
        android:layout_rowWeight="1"
        android:text="1"/>
```

```xml
<Button
    android:layout_columnWeight="1"
    android:layout_rowWeight="1"
    android:text="2"/>
<Button
    android:layout_columnWeight="1"
    android:layout_rowWeight="1"
    android:text="3"/>
<Button
    android:layout_columnWeight="1"
    android:layout_rowWeight="1"
    android:text="-"/>
<Button
    android:layout_columnWeight="1"
    android:layout_rowWeight="1"
    android:text="4"/>
<Button
    android:layout_columnWeight="1"
    android:layout_rowWeight="1"
    android:text="5"/>
<Button
    android:layout_columnWeight="1"
    android:layout_rowWeight="1"
    android:text="6"/>
<Button
    android:layout_columnWeight="1"
    android:layout_rowWeight="1"
    android:text="+"/>
<Button
    android:layout_columnWeight="1"
    android:layout_rowWeight="1"
    android:text="7"/>
<Button
    android:layout_columnWeight="1"
    android:layout_rowWeight="1"
    android:text="8"/>
<Button
    android:layout_columnWeight="1"
    android:layout_rowWeight="1"
```

```
        android:text="9"/>
    <Button
        android:layout_columnWeight="1"
        android:layout_rowWeight="1"
        android:layout_rowSpan="2"
        android:text="="
        android:background="#228B22"/>
    <Button
        android:layout_columnWeight="1"
        android:layout_rowWeight="1"
        android:layout_columnSpan="2"
        android:text="0"/>
    <Button
        android:layout_columnWeight="1"
        android:layout_rowWeight="1"
        android:text="."/>
</GridLayout>
```

(3) 打开 java/cc.turbosnail.uidemo/MainActivity,修改代码如下:

```
protected void onCreate(Bundle savedInstanceState) {
    super.onCreate(savedInstanceState);
    //setContentView(R.layout.activity_main);
    setContentView(R.layout.ly_gridlayout);//设置 ly_gridlayout 为当前显示视图
}
```

程序运行效果如图 2.41 所示。

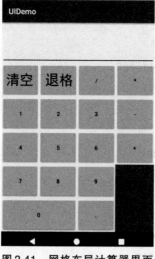

图 2.41　网格布局计算器界面

本例通过权重值"1"来平分行和列,当需要完全通过权重来划分空间时,组件的对应宽度(android:layout_width)或者高度(android:layout_height)属性应该设置为"0"或者不设置,这样才能达到划分所有空间的效果。如果设置了组件的宽度(或者高度),则权重只会划分当前行总宽度(或者列总高度)减去当前行(或者列)上所有组件宽度(或者高度)之差后的剩余空间,按权重比例加给各个组件。

### 2.3.6 绝对布局(AbsoluteLayout)

绝对布局以像素坐标的形式指定UI组件在布局内的位置,每个UI组件都需要通过X坐标值和Y坐标值来定位,(0,0)坐标为布局的左上角位置。在UI组件的属性中通过android:layout_x、android:layout_y,以dp为单位确定位置,代码如下:

```
<AbsoluteLayout xmlns:android="http://schemas.android.com/apk/res/android"
    android:layout_width="match_parent" android:layout_height="match_parent">
    <Button
        android:layout_width="wrap_content"
        android:layout_height="wrap_content"
        android:layout_x="200dp"
        android:layout_y="60dp"
        android:text="Button"/>
</AbsoluteLayout>
```

在移动终端屏幕分辨率多样的现状下,绝对布局无法适应,所以在开发中不推荐使用。

### 2.3.7 约束布局(ConstraintLayout)

约束布局通过为组件建立相对于其他组件的位置约束关系来进行布局,约束关系包括上、下、左、右的任何位置及距离。

约束布局是Android Studio 2.3开始新建项目默认的布局方式,相对于其他布局,它具有明显优势:

(1)提供了比相对布局更为细致的定位关系,可以更灵活地进行复杂界面的布局。

(2)通过单一层面的约束关系定位所有组件,可以有效地解决界面实现中布局嵌套过多的问题,提高程序运行效率。

(3)相对于其他布局方式,约束布局比较适合使用可视化的方式来编写界面,不用过多关注XML代码,提高程序开发效率。

相对于直接编写XML代码的界面实现方式,约束布局更适合通过可视化的方式来实现。在Android studio中通过拖动控件的方式来布局,然后进行细节调整。

[例2.19] 约束布局简单示例。在项目"UIDemo"中,首先自定义一个基于约束布局的XML布局文件,然后设置该文件为当前显示视图。

(1)在res/layout目录上点击右键,选择"New"→"Layout Resource file",填写"File name"为"ly_constraintlayout","Root element"为默认的"android.support.constraint.ConstraintLayout",

如图 2.42 所示。

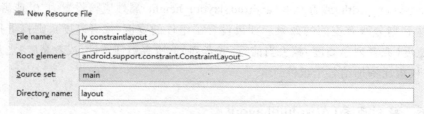

图 2.42　创建约束布局文件

（2）布局新建后打开代码编辑窗口左下方的"Design"设计视图,此时可以看到可视化的布局编辑窗口,如图 2.43 所示。

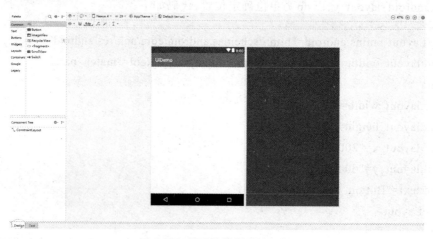

图 2.43　代码编辑器设计视图

（3）从 Palette 菜单中拖出一个 TextView 组件至布局编辑窗口的手机界面上,点击该组件四周的圆点处,可以看到白色的圆点被放大成闪烁的绿色圆点,如图 2.44 所示。

（4）沿着绿色圆点向上拖动鼠标,直到界面顶端,此时代表着这个 TextView 组件和布局顶部建立了约束关系,如图 2.45 所示。

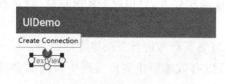

图 2.44　添加组件到约束布局

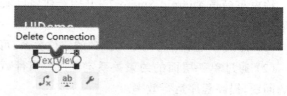

图 2.45　添加上约束

（5）拖动下方的小绿色圆点到界面底端,建立这个 TextView 组件和布局底部的约束关系,此时可以上下调整组件的位置,如图 2.46 所示。

（6）同理,建立这个组件和布局左、右两边的约束关系,此时这个组件就有了上、下、左、右四个方向的约束,即完成了定位,如图 2.47 所示。这时鼠标指向白色圆点,可以看到,圆点变成了红色。如果要删除某一个约束关系,则单击一下对应的红点,红点变成绿色,约束关系解除,如图 2.48 所示。

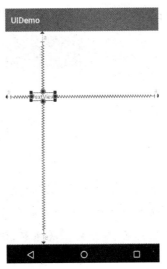

图2.46　添加下约束　　　　　图2.47　添加左、右约束

同时，布局编辑窗口右边是组件的属性编辑窗口，可以在这里修改组件的属性，例如id、layout_width、margin、text等，如图2.49所示。

（7）再从Palette菜单中拖出一个EditText组件（Plain text）至布局编辑窗口，并添加上、下、左、右约束，其中左侧约束添加至TextView的右侧，这样建立组件之间的约束关系，如图2.50所示。然后在右侧属性编辑窗口修改id属性为"edt_name"，"hint"属性为"请输入用户名"。

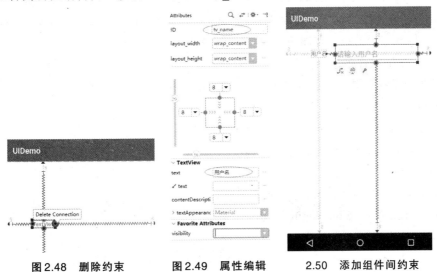

图2.48　删除约束　　　图2.49　属性编辑　　　2.50　添加组件间约束

（8）打开java/cc.turbosnail.uidemo/MainActivity，修改代码如下：

```
protected void onCreate(Bundle savedInstanceState) {
    super.onCreate(savedInstanceState);
    //setContentView(R.layout.activity_main),
    setContentView(R.layout.ly_constraintlayout);//设置ly_constraintlayout为当前显示视图
}
```

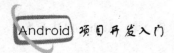

程序运行效果如图2.51所示。

图2.51 约束布局运行效果

说明：

- 在设置约束关系时，应该同时至少在互相垂直的方向设置2个约束，即至少一个水平方向和一个竖直方向需要有约束。
- 拖入组件后、建立约束关系前，请先根据编码规范修改组件的id属性。可以在属性编辑窗口直接修改，也可打开编辑器的Text视图修改。
- 未建立任何约束关系的组件，即使在设计视图中被放到手机界面的某个位置，但是在运行时也会被默认放置到手机界面的左上角位置。

[项目案例2] 设计并实现个人记账本项目的用户登录界面

（1）在项目"AccountBook"中，自定义一个基于约束布局的XML布局文件。

在res/layout目录上点击右键，选择"New"→"Layout Resource file"，填写"File name"为"activity_login"，"Root element"为默认的"android.support.constraint.ConstraintLayout"。

（2）打开布局的设计视图窗口，建立约束关系如图2.52所示。

说明：在把组件拖出到设计视图上时，请先为组件自定义符合编码规范的id属性，再设置边距文本等各类属性以及建立约束关系，例如本例中的id属性如下：

android:id="@+id/iv_logo_login"

android:id="@+id/edt_username_login"

android:id="@+id/edt_password_login"

android:id="@+id/ck_remeber_login"

android:id="@+id/ck_music_login"

android:id="@+id/btn_login_login"

android:id="@+id/tv_register_login"

（3）打开java/cc.turbosnail.accountbook/MainActivity，修改代码如下：

```
protected void onCreate(Bundle savedInstanceState) {
    super.onCreate(savedInstanceState);
    //setContentView(R.layout.activity_main);
    setContentView(R.layout.activity_login);//设置activity_login为当前显示视图
}
```

程序运行效果如图2.53所示。

## 第 2 章 Android UI 开发

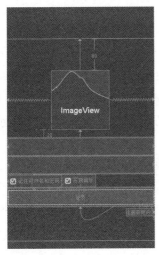

图 2.52 登录界面约束关系图

2.53 登录界面运行效果图

在代码编辑窗口中打开"Text"视图，可以看到为约束布局自动生成的布局代码，这些代码包括组件自身属性设置和约束关系的定义，在开发中可以通过直接修改这些代码让布局更细致。约束关系常见的属性如表 2.18 所示。

表 2.18 约束布局常见属性

| 属性声明 | 功能描述 |
| --- | --- |
| app:layout_constraintLeft_toLeftOf | 表示此控件的左边框与某个控件的左边框对齐 |
| app:layout_constraintLeft_toRightOf | 表示此控件的左边框与某个控件的右边框对齐 |
| app:layout_constraintRight_toLeftOf | 表示此控件的右边框与某个控件的左边框对齐 |
| app:layout_constraintRight_toRightOf | 表示此控件的右边框与某个控件的右边框对齐 |
| app:layout_constraintTop_toTopOf | 表示此控件的上边框与某个控件的上边框对齐 |
| app:layout_constraintTop_toBottomOf | 表示此控件的上边框与某个控件的下边框对齐 |
| app:layout_constraintBottom_toTopOf | 表示此控件的下边框与某个控件的上边框对齐 |
| app:layout_constraintBottom_toBottomOf | 表示此控件的下边框与某个控件的下边框对齐 |
| app:layout_constraintStart_toStartOf | 表示此控件的起始位置与某个控件的起始位置对齐 |
| app:layout_constraintStart_toEndOf | 表示此控件的起始位置与某个控件的结束位置对齐 |
| app:layout_constraintEnd_toStartOf | 表示此控件的结束位置与某个控件的起始位置对齐 |
| app:layout_constraintEnd_toEndOf | 表示此控件的结束位置与某个控件的结束位置对齐 |
| app:layout_constraintBaseline_toBaselineOf | 表示此控件与某个控件的文本基准线对齐 |
| app:layout_constraintHorizontal_bias | 调整水平方向的偏移比例，取值[0,1]，需要先设置 start 和 end 属性 |
| app:layout_constraintVertical_bias | 调整垂直方向的偏移比例，取值[0,1]，需要先设置 top 和 bottom 属性 |

## 2.4 布局嵌套

布局作为容器使用,它本身也是UI组件,所以可以把一个布局放置到另一个布局中,实现布局的嵌套。通过布局嵌套,可以更多样化地实现界面效果,但是在应用开发中布局嵌套的层次不宜太深。在可以通过扁平化的界面布局方式如约束布局、相对布局直接实现的情况下,建议不使用布局嵌套。

[例2.20] 布局嵌套简单示例。在项目"UIDemo"中,首先自定义一个基于相对布局的XML布局文件,然后设置该文件为当前显示视图。

(1) 在res/layout目录上点击右键,选择"New"→"Layout Resource file",填写"File name"为"ly_nested","Root element"为默认的"RelativeLayout",如图2.54所示。

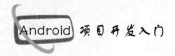

图2.54 新建布局文件ly_nested.xml

(2) 编写布局代码。打开res/layout/ly_nested.xml文件,切换到"text"视图模式,编写如下布局代码:

```
<?xml version="1.0" encoding="utf-8"?>
<RelativeLayout xmlns:android="http://schemas.android.com/apk/res/android"
    android:layout_width="match_parent"
    android:layout_height="match_parent">
    <LinearLayout
        android:id="@+id/ly_top"
        android:layout_width="match_parent"
        android:layout_height="wrap_content"
        android:orientation="vertical">
        <Button
            android:layout_width="wrap_content"
            android:layout_height="wrap_content"
            android:text="按钮"/>
        <Button
            android:layout_width="wrap_content"
            android:layout_height="wrap_content"
            android:text="按钮"/>
        <Button
            android:layout_width="wrap_content"
```

　　　　android:layout_height="wrap_content"
　　　　android:text="按钮"/>
　</LinearLayout>
　<LinearLayout
　　　android:layout_width="match_parent"
　　　android:layout_height="wrap_content"
　　　android:layout_below="@+id/ly_top"
　　　android:background="@color/colorPrimaryDark"
　　　android:orientation="horizontal">
　　　<Button
　　　　android:layout_width="wrap_content"
　　　　android:layout_height="wrap_content"
　　　　android:text="按钮"/>
　　　<Button
　　　　android:layout_width="wrap_content"
　　　　android:layout_height="wrap_content"
　　　　android:text="按钮"/>
　　　<Button
　　　　android:layout_width="wrap_content"
　　　　android:layout_height="wrap_content"
　　　　android:text="按钮"/>
　</LinearLayout>
</RelativeLayout>

（3）打开java/cc.turbosnail.uidemo/MainActivity，修改代码如下：

```java
protected void onCreate(Bundle savedInstanceState) {
    super.onCreate(savedInstanceState);
    //setContentView(R.layout.activity_main);
    setContentView(R.layout.ly_gridlayout);//设置ly_gridlayout为当前显示视图
}
```

程序运行效果如图2.55所示。

图2.55　布局嵌套运行效果图

在本例中，根布局为相对布局，在其中添加了两个线性布局，第二个线性布局通过 android:layout_below 属性被放置在第一个线性布局的下方，两个线性布局分别为垂直和水平的方式添加了 3 个按钮，第二个线性布局还设置了背景颜色。

## 2.5　UI 组件在 Java 代码和 XML 文件中调用

### 2.5.1　在 Java 代码中访问 XML 文件中的组件

通过 XML 文件的形式设计出软件界面后，在程序中可以通过 id 属性获取和访问对应的组件，并可以进一步对组件的外观进行设置、获取组件中的输入值、响应组件上发生的事件等。

UI 组件都有与它同名的一个 Java 类，Android 提供 findViewById 方法来从当前布局文件中查找和获取 UI 组件并生成对应的 Java 对象，该方法在 Activity 中的声明如下：

public View findViewById(int id)，通过 XML 文件中的 id 属性获取一个 View 对象。因为返回的是 View 类型，所以需要通过强制类型转换的方式把它指向对应子类型的变量。示例如下：

```
Button btnLogin = (Button) findViewById(R.id.btn_login);
EditText edtUsername = (EditText) findViewById(R.id.edt_username_login);
TextView tvRegister = (TextView) findViewById(R.id.tv_register_login);
CheckBox cbRemeber = (CheckBox) findViewById(R.id.ck_remeber);
```

从 Android Studio 3 开始，可以直接写为：

```
Button btnLogin = findViewById(R.id.btn_login);
EditText edtUsername = findViewById(R.id.edt_username_login);
TextView tvRegister = findViewById(R.id.tv_register_login);
CheckBox cbRemeber = findViewById(R.id.ck_remeber);
```

[例 2.21] 在 Java 代码中获取 UI 组件对象。

（1）在项目"UIDemo"中新建一个布局文件 ly_getview.xml，布局界面如图 2.56 所示。

图 2.56　ly_getview 布局效果

其中，编辑框和按钮的 id 属性设置代码如下：

```xml
<EditText
    android:id="@+id/edt_input"
    ……
/>
<Button
    android:id="@+id/btn_submit"
    ……
/>
```

（2）打开Java代码java/cc.turbosnail.uidemo/MainActivity，设置ly_getview.xml为当前显示视图，然后根据id获取到布局中的编辑框和按钮，并修改属性如下：

```java
package cc.turbosnail.uidemo;
import android.support.v7.app.AppCompatActivity;
import android.os.Bundle;
import android.widget.Button;
import android.widget.EditText;
public class MainActivity extends AppCompatActivity {
    private EditText edtInput;
    private Button btnSubmit;
    @Override
    protected void onCreate(Bundle savedInstanceState) {
        super.onCreate(savedInstanceState);
        setContentView(R.layout.ly_getview);//设置ly_getview为当前显示视图
        edtInput=findViewById(R.id.edt_input);//通过id获取编辑框对象
        btnSubmit=findViewById(R.id.btn_submit);//通过id获取按钮对象
        /*修改属性*/
        edtInput.setText("新的输入");
        btnSubmit.setText("已提交");
        btnSubmit.setBackgroundColor(0xFF6E6E6E);//(透明度\红色\绿色\蓝色)
    }
}
```

运行程序，可以看到界面UI组件显示外观效果是Java代码中修改后的效果，如图2.57所示。

图2.57　在Java代码中修改属性

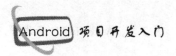

### 2.5.2 在Java代码中创建和使用组件

Android也允许在Java代码中创建布局和组件的对象,通过位置、嵌套关系的设置来实现软件界面,例如:

```
protected void onCreate(Bundle savedInstanceState) {
    super.onCreate(savedInstanceState);
    RelativeLayout rootLayout = new RelativeLayout(this);//创建布局对象
    rootLayout.setBackgroundColor(Color.rgb(0, 0, 255));//设置背景颜色
    Button button = new Button(this);//创建按钮对象
    RelativeLayout.LayoutParams btnParams = new RelativeLayout.LayoutParams(
            RelativeLayout.LayoutParams.WRAP_CONTENT,
            RelativeLayout.LayoutParams.WRAP_CONTENT);//创建布局规则
    btnParams.addRule(RelativeLayout.CENTER_HORIZONTAL);//添加布局规则
    button.setLayoutParams(btnParams);//设置按钮布局属性
    button.setText("按钮");//设置按钮自身属性
    rootLayout.addView(button);//为布局添加组件
    setContentView(rootLayout);//设置linearLayout为当前显示视图
}
```

这种用Java代码来实现界面的方法工作量大,不宜修改,界面和逻辑难以分离,在开发中一般不推荐使用。但是对于一些需要临时生成的界面组件来说,这种方式可以让界面变化更加灵活,也具有了自身的应用价值。

## 2.6 列表(ListView)与适配器

列表(ListView)用来以分行的形式显示多个结构相同的数据项,每个数据项被称为一个列表项,在列表项数量多至超出屏幕的情况下,可以通过上下滑动操作来进行显示。在Android中,列表应用非常广泛,如通信录列表、短信列表、数据列表等均可以通过列表来实现。

当列表项数量多,超出屏幕显示范围时,未显示的列表项不会被进行绘制,而是被放置在一个缓存区,根据用户的滑动操作,从缓存区读取内容进行屏幕绘制,这样的设计方式可以节省系统资源。列表的常用属性如表2.19所示。

表2.19 列表常用属性

| 属性声明 | 功能描述 | 属性值 |
| --- | --- | --- |
| android:listSelector | 列表项在选中时的显示效果 | 颜色或者图片均可,如"#0F0",此时选中时显示为绿色 |
| android:divider | 列表项分割线的效果 | 颜色或者图片均可,如"#00F" |
| android:dividerHeight | 列表项分割线的高度 | 数值型,如"5 dp" |
| android:scrollbars | 设置滚动条的状态 | "vertical"、"horizontal"或者"none" |

适配器(Adapter)在Android中大量存在,它用来连接数据源和UI组件,形成的关系如图2.58所示。

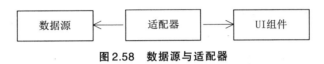

图2.58 数据源与适配器

适配器对数据源进行整理和封装,然后按统一的接口根据需要传递给UI组件显示。这种模式可以有效地降低数据源和UI组件之间的耦合,特别是对于一些数据源结构复杂、数据量大的情况,可以有效地提升灵活性和程序效率。常见的适配器有BaseAdapter、ArrayAdapter、SimpleAdapter等。

☞ ArrayAdapter:最简单的适配器,每个列表项只能是一行文字,支持泛型操作。

☞ SimpleAdapter:具有良好扩展性的适配器,列表项可以定义各种布局样式,且能放置各种UI组件,如图片对象视图、按钮、复选框等。

☞ BaseAdapter:是以上两种适配器的父类,它可以用来自己定义适配器,能将各种复杂组合的数据源与自定义的布局连接起来,以实现用户需要的展示效果,具有很强的扩展性。

列表和它的数据源通过适配器来连接。其实现过程如下:

(1)设计并实现列表中一个列表项,即一个条目的界面布局。一般是一个独立的xml布局文件。

(2)选择合适的适配器,根据适配器的需要构建数据源。ArrayAdapter的数据源一般为简单的字符串数组,SimpleAdapter的数据源可以是比较复杂的数据结构,如Map型链表(List),BaseAdapter的数据源可以是更为复杂数据结构,如由自定义类构成的链表(List)。

(3)创建适配器类的对象。该对象负责关联布局文件和数据源。如果是BaseAdapter,需要开发者自定义子类,并重写四个方法。

(4)获取主布局中的ListView对象,并为其设置适配器。

例如,在ListView中一个典型的ArrayAdapte适配器用法如下:

```
ListView list = findViewById(R.id.lv_contacts);
//定义字符串数组作为数据源
String [] arrayData = {"狐狸","玫瑰","国王"};
```

//新建适配器对象绑定数据,参数列表为(当前Activity,android自带的列表项布局文件之一,数据源)
ArrayAdapter adapter = new ArrayAdapter<String>(this, android.R.layout.simple_list_item_1, arrayData);
//视图(ListView)加载适配器
list.setAdapter(adapter);
如果是自定义一个布局文件item_a.xml,代码如下:
```xml
<?xml version="1.0" encoding="utf-8"?>
<LinearLayout xmlns:android="http://schemas.android.com/apk/res/android"
    android:orientation="vertical" android:layout_width="match_parent"
    android:layout_height="match_parent">
    <TextView
        android:id="@+id/tv_info"
        android:layout_width="match_parent"
        android:layout_height="wrap_content"
        android:text="item"
        />
</LinearLayout>
```
则可以使用ArrayAdapter的另一个构造方法如下:
```
ListView list = findViewById(R.id.lv_contacts);
String[] arrayData = {"狐狸","玫瑰","国王"};
ArrayAdapter adapter = new ArrayAdapter(MainActivity.this,R.layout.item_a,R.id.tv_info,arrayData);
list.setAdapter(adapter);
```

再例如,在ListView中一个典型的SimpleAdapter适配器用法如下:

```
private String[] name = {"狐狸","玫瑰","国王"};
private String[] info ={"聪明的狐狸","带刺的花儿","没有一个臣民"};
private int[] image = {R.mipmap.ic_launcher,R.mipmap.ic_launcher,R.mipmap.ic_launcher};
List<Map<String,Object>> list=new ArrayList<>();
for(int i=0;i<name.length;i++){
    Map<String, Object> map=new HashMap<>();
    map.put("image",image[i]);
    map.put("name",name[i]);
    map.put("info",image[i]);
    lists.add(map);
}
SimpleAdapter adapter=new SimpleAdapter(MainActivity.this,list, R.layout.item_s,new String[]
```

{"image","name","info"},new int[]{R.id.iv_photo,R.id.tv_name,R.id.tv_info});
listView.setAdapter(adapter);

关于本例中SimpleAdapter的说明：

（1）R.layout.item_s是一个自定义的简单布局文件item_s.xml，其中含有1个ImageView、2个TextView组件，其id属性分别为：iv_photo、tv_name、tv_info。

（2）new String[ ]和new int[ ]中的参数分别为Map中的key值和xml布局文件中的组件id，它们要一一对应。

[例2.21] 基于ListView和BaseAdapter适配器，创建一个通信录的显示界面。

（1）新建一个Android项目，名称为ListViewDemo。

（2）新建一个表示列表项的布局文件。在res/layout目录上点击右键，新建"XML"→"Layout XML File"，并命名为"item_contacts"，布局为"LinearLayout"，如图2.59所示。

图2.59 创建列表项布局文件

在item_contacts.xml中为列表设计单个列表项的布局样式，代码如下：

```xml
<?xml version="1.0" encoding="utf-8"?>
<LinearLayout xmlns:android="http://schemas.android.com/apk/res/android"
    android:layout_width="match_parent"
    android:layout_height="match_parent"
    android:orientation="horizontal">
    <ImageView
        android:id="@+id/iv_photo"
        android:layout_width="48dp"
        android:layout_height="48dp"
        android:layout_marginLeft="10dp"
        android:src="@mipmap/ic_launcher"/>
    <LinearLayout
        android:layout_width="match_parent"
        android:layout_height="48dp"
        android:layout_marginLeft="10dp"
        android:orientation="vertical">
        <TextView
            android:id="@+id/tv_name"
            android:layout_width="match_parent"
```

```xml
            android:layout_height="wrap_content"
            android:gravity="center|left"
            android:text="姓名"
            android:textColor="#000000"
            android:textSize="16sp"/>
        <TextView
            android:id="@+id/tv_info"
            android:layout_width="match_parent"
            android:layout_height="wrap_content"
            android:gravity="center|left"
            android:text="描述信息"
            android:textColor="#999999"
            android:textSize="14sp"/>
    </LinearLayout>
</LinearLayout>
```

（3）新建代表一个列表项所需的实体类。在项目目录"java"→"cc.turbosnail.listviewdemo"上点击右键，选择"New"→"Java Class"，创建一个Java类，命名为ContactsBean，代码如下：

```java
package cc.turbosnail.listviewdemo;
public class ContactsBean {
    private int imgId;//图像资源ID
    private String name;
    private String info;
    public ContactsBean(int imgId, String name, String info) {
        this.imgId = imgId;
        this.name = name;
        this.info = info;
    }
    public int getImgId() {
        return imgId;
    }
    public String getName() {
        return name;
    }
    public String getInfo() {
        return info;
    }
}
```

(4) 创建适配器。在项目目录 java→cc.turbosnail.listviewdemo 上点击右键,选择 New→Java Class,创建一个 Java 类,命名为 ContactsAdapter,代码如下:

```java
package cc.turbosnail.listviewdemo;
import android.content.Context;
import android.view.LayoutInflater;
import android.view.View;
import android.view.ViewGroup;
import android.widget.BaseAdapter;
import android.widget.ImageView;
import android.widget.TextView;
import java.util.List;
public class ContactsAdapter extends BaseAdapter {
    private List<ContactsBean> contactsList;
    private LayoutInflater inflater;
    //通过构造方法关联数据源与数据适配器
    public ContactsAdapter(Context context, List<ContactsBean> list) {
        contactsList = list;
        inflater = LayoutInflater.from(context);
    }
    @Override
    public int getCount() {
        //return 0;
        return contactsList.size();
    }
    @Override
    public Object getItem(int position) {
        //return null;
        return contactsList.get(position);
    }
    @Override
    public long getItemId(int position) {
        //return 0;
        return position;
    }
    @Override
    public View getView(int position, View convertView, ViewGroup parent) {
        //return null;
        ViewHolder viewHolder;
```

```java
        //如果view未被实例化过,缓存池中没有对应的缓存
        if (convertView == null) {
            viewHolder = new ViewHolder();
            convertView = inflater.inflate(R.layout.item_contacts, null);
            viewHolder.imageView = convertView.findViewById(R.id.iv_photo);
            viewHolder.title = convertView.findViewById(R.id.tv_name);
            viewHolder.content = convertView.findViewById(R.id.tv_info);
            /*为convertView添加一个附件信息,setTag方法的参数是Object型*/
            convertView.setTag(viewHolder);
        }else{
            viewHolder = (ViewHolder) convertView.getTag();
        }
        ContactsBean bean = contactsList.get(position);
        viewHolder.imageView.setImageResource(bean.getImgId());
        viewHolder.title.setText(bean.getName());
        viewHolder.content.setText(bean.getInfo());
        return convertView;
    }
    class ViewHolder{
        public ImageView imageView;
        public TextView title;
        public TextView content;
    }
}
```

在本段代码中,通过一个ViewHolder内部类,将当前convertView对象的内部组件对象存储起来,并通过UI组件的setTag方法设置成一个附加信息,在实际显示时,如果这个convertView对象已经存在,则直接从ViewHolder对象的附加信息中通过getTag方法读取界面组件对象,然后赋值。这样做就避免了每次通过findViewById获取组件对象造成的性能损耗,特别是组件比较多的时候,这种写法有利于提高程序的运行效率。下面列出每次重复调用findViewById方法的代码段,请读者对比分析。

```java
public View getView(int position, View convertView, ViewGroup parent) {
    if (convertView == null) {
        convertView = inflater.inflate(R.layout.item_contacts, null);
    }
    ImageView imageView = convertView.findViewById(R.id.iv_photo);
    TextView title = convertView.findViewById(R.id.tv_name);
    TextView content = convertView.findViewById(R.id.tv_info);
```

```
        ContactsBean bean = contactsList.get(position);
        imageView.setImageResource(bean.getImgId());
        title.setText(bean.getName());
        content.setText(bean.getInfo());
        return convertView;
    }
}
```

(5)在界面上添加 ListView 组件,并在 Activity 代码中获取该组件,通过适配器传递数据。

activity_main.xml 代码清单:

```xml
<?xml version="1.0" encoding="utf-8"?>
<android.support.constraint.ConstraintLayout xmlns:android="http://schemas.android.com/apk/res/android"
    xmlns:app="http://schemas.android.com/apk/res-auto"
    xmlns:tools="http://schemas.android.com/tools"
    android:layout_width="match_parent"
    android:layout_height="match_parent"
    tools:context="cc.turbosnail.listviewdemo.MainActivity">
    <ListView
        android:id="@+id/lv_contacts"
        android:layout_width="match_parent"
        android:layout_height="match_parent"
        android:listSelector ="#317EF3"
        app:layout_constraintTop_toTopOf="parent"
        app:layout_constraintRight_toRightOf="parent"
        app:layout_constraintLeft_toLeftOf="parent"
        app:layout_constraintBottom_toBottomOf="parent"/>
</android.support.constraint.ConstraintLayout>
```

MainActivity.java 代码清单:

```java
package cc.turbosnail.listviewdemo;
import android.support.v7.app.AppCompatActivity;
import android.os.Bundle;
import android.widget.ListView;
import java.util.ArrayList;
import java.util.List;
public class MainActivity extends AppCompatActivity {
    ListView listView;
```

```
@Override
protected void onCreate(Bundle savedInstanceState) {
    super.onCreate(savedInstanceState);
    setContentView(R.layout.activity_main);
    listView = findViewById(R.id.lv_contacts);
    List<ContactsBean> contactsBeanList = new ArrayList<>();
    contactsBeanList.add(new ContactsBean(R.mipmap.ic_launcher, "狐狸", "聪明的狐狸"));
    contactsBeanList.add(new ContactsBean(R.mipmap.ic_launcher, "玫瑰", "带刺的花儿"));
    contactsBeanList.add(new ContactsBean(R.mipmap.ic_launcher, "国王", "没有一个臣民"));
    listView.setAdapter(new ContactsAdapter(this,contactsBeanList));
}
```

程序运行效果如图2.60所示。

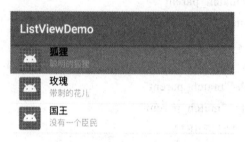

图2.60 列表(ListView)示例

## 2.7 简洁灵活的列表(RecyclerView)

RecylerView是support:recyclerview-v7包中提供的用于展示大量数据的滑动UI组件。它与ListView具有相似之处，在很多场景下可以代替ListView。RecyclerView本身封装了ViewHolder，简化了适配过程，代码逻辑更加简洁。同时，它对列表显示与数据加载进行了解耦，提供了通过布局管理器控制布局效果、自定义分割线和增删动画等功能，具有非常强的扩展性，程序实现变得更加灵活。在项目开发中，列表显示优先推荐使用RecylerView。

RecylerView的一般实现流程如下：
(1) 设计并实现列表中的一个列表项，即一个条目的界面布局。
(2) 在项目目录build.gradle(Module:app)文件中添加依赖导入support-v7包。
(3) 创建适配器类。

首先导入包：

```
import android.support.v7.widget.RecyclerView;//导入包
```

然后声明该类继承 RecyclerView.Adapter<VH>，VH 是一个自定义的继承了 RecyclerView.ViewHolder 的内部类，用来完成组件对象的初始化工作。例如：

```
class ViewHolder extends RecyclerView.ViewHolder {
    ImageView ivPhoto;
    TextView tvName, tvInfo;
    public ViewHolder(View itemView) {
        super(itemView);
        ivPhoto = itemView.findViewById(R.id.iv_photo);
        tvName = itemView.findViewById(R.id.tv_name);
        tvInfo = itemView.findViewById(R.id.tv_info);
    }
}
```

同时，在适配器的构造方法中添加一个获取数据的参数，例如自定义的数据列表 List<ContactsBean> 用来向适配器传递数据项：

```
public ContactsAdapter(Context context, List<ContactsBean> list) {
    this.context = context;
    this.list = list;
}
```

接着，要实现3个方法：

- public ContactsAdapter.ViewHolder onCreateViewHolder(ViewGroup viewGroup, int i)，该方法为每个列表项生成一个 view 对象出来，它返回自定义 VH 型的 ViewHolder 对象，实现对 View 的封装。一个典型的实现如下：

```
public ContactsAdapter.ViewHolder onCreateViewHolder(ViewGroup viewGroup, int i) {
    View view = LayoutInflater.from(viewGroup.getContext()).inflate(R.layout.item_contacts, viewGroup, false);
    ViewHolder viewHolder = new ViewHolder(view);
    return viewHolder;
    // return null;
}
```

- public void onBindViewHolder(ContactsAdapter.ViewHolder viewHolder, int i)，该方法用来把数据添加到 View 中去，对于组件中的事件处理，也可以放在这个方法中。一个典型的实现如下：

```
public void onBindViewHolder(ContactsAdapter.ViewHolder viewHolder, int i) {
    /*把数据添加到对应组件中*/
```

```
            ContactsBean bean = list.get(i);
            viewHolder.ivPhoto.setImageResource(bean.getImgId());
            viewHolder.tvName.setText(bean.getName());
            viewHolder.tvInfo.setText(bean.getInfo());
            viewHolder.tvName.setOnClickListener(new View.OnClickListener() {
                @Override
                public void onClick(View v) {
                    //item 点击事件处理
                }
            });
        }
```

☞ public int getItemCount(·),该方法用来返回数据的总数目。例如,数据从List中获取,那么该方法可以写为:

```
        public int getItemCount() {
            return list.size();
            //return 0;
        }
```

(4) 在界面布局中添加 RecyclerView 组件,在 Activity 代码中获取该组件,通过适配器传递数据,并设置列表布局以及分割线等属性。其中,通过 Layout Manager 布局管理器为列表设置布局是必需的,Android 通过提供线性布局管理器(LinearLayoutManager)、网格布局管理器(GridLayoutManager)、瀑布流布局管理器(StaggeredGridLayoutManager)来灵活展示列表内容。例如,线性布局的设置如下:

```
LinearLayoutManager layout = new LinearLayoutManager(this);
//设置布局管理器
recyclerView.setLayoutManager(layout);
```

[例2.22] 基于RecylerView,实现一个通信录的显示界面。

(1) 新建一个Android项目,名称为RecylerViewDemo。

(2) 新建一个表示列表项的布局文件。在 res/layout 目录上点击右键,新建"XML"→"Layout XML File",并命名为"item_contacts"。布局为"LinearLayout",布局代码与例2.13 ListView列表项相同,列表项根布局的高度不能是"match_parent",需要手动确定,例如:
android:layout_height="50dp"。

(3) 新建代表一个列表项所需的实体类。在项目目录"java"→"cc.turbosnail.recyclerviewdemo"上点击右键,选择"New"→"Java Class",创建一个Java类,命名为ContactsBean,代码与例2.13 ListView 的实体类ContactsBean相同。

(4) 创建适配器。在项目目录"java"→"cc.turbosnail.recyclerviewdemo"上点击右键,选

择"New"→"Java Class",创建一个Java类,命名为ContactsAdapter。

在项目目录"Gradle Scripts"→"build.gradle(Module:app)"中添加"implementation 'com.android.support:recyclerview-v7:28.0.0'",版本号"28.0.0"是当前项目中应用的版本号,可以参照dependencies中已有的语句填写,版本号必须一致。

然后在适配器Java代码中导入包,实现代码如下:

```java
package cc.turbosnail.recyclerviewdemo;
import android.content.Context;
import android.support.v7.widget.RecyclerView;//导入包
import android.view.LayoutInflater;
import android.view.View;
import android.view.ViewGroup;
import android.widget.ImageView;
import android.widget.TextView;
import java.util.List;
public class ContactsAdapter extends RecyclerView.Adapter<ContactsAdapter.ViewHolder> {
    private Context context;
    private List<ContactsBean> list;
    public ContactsAdapter(Context context, List<ContactsBean> list) {
        this.context = context;
        this.list = list;
    }
    @Override
    public ContactsAdapter.ViewHolder onCreateViewHolder(ViewGroup viewGroup, int i) {
        View view = LayoutInflater.from(viewGroup.getContext()).inflate(R.layout.item_contacts, viewGroup, false);
        ViewHolder viewHolder = new ViewHolder(view);
        return viewHolder;
        //return null;
    }
    @Override
    public void onBindViewHolder(ContactsAdapter.ViewHolder viewHolder, int i) {
        ContactsBean bean = list.get(i);
        viewHolder.ivPhoto.setImageResource(bean.getImgId());
        viewHolder.tvName.setText(bean.getName());
        viewHolder.tvInfo.setText(bean.getInfo());
    }
    @Override
    public int getItemCount() {
```

```java
            return list.size();
            //return 0;
        }
        public class ViewHolder extends RecyclerView.ViewHolder {
            ImageView ivPhoto;
            TextView tvName, tvInfo;
            public ViewHolder(View itemView) {
                super(itemView);
                ivPhoto = itemView.findViewById(R.id.iv_photo);
                tvName = itemView.findViewById(R.id.tv_name);
                tvInfo = itemView.findViewById(R.id.tv_info);
            }
        }
    }
```

(5) 在界面上添加RecyclerView组件,并在Activity代码中获取该组件,通过适配器传递数据,并设置列表布局以及分割线等属性。

activity_main.xml代码清单:

```xml
<?xml version="1.0" encoding="utf-8"?>
<android.support.constraint.ConstraintLayout xmlns:android="http://schemas.android.com/apk/res/android"
    xmlns:app="http://schemas.android.com/apk/res-auto"
    xmlns:tools="http://schemas.android.com/tools"
    android:layout_width="match_parent"
    android:layout_height="match_parent"
    tools:context=".MainActivity">
    <android.support.v7.widget.RecyclerView
        android:id = "@+id/rcl_contacts"
        android:layout_width="match_parent"
        android:layout_height="match_parent"/>
</android.support.constraint.ConstraintLayout>
```

MainActivity.java代码清单:

```java
package cc.turbosnail.recyclerviewdemo;
import android.support.v7.app.AppCompatActivity;
import android.os.Bundle;
import android.support.v7.widget.DefaultItemAnimator;
import android.support.v7.widget.DividerItemDecoration;
```

```java
import android.support.v7.widget.LinearLayoutManager;
import android.support.v7.widget.RecyclerView;
import java.util.ArrayList;
import java.util.List;
public class MainActivity extends AppCompatActivity {
    private List<ContactsBean> list = new ArrayList<>();
    private RecyclerView recyclerView;
    private ContactsAdapter adapter;
    @Override
    protected void onCreate(Bundle savedInstanceState) {
        super.onCreate(savedInstanceState);
        setContentView(R.layout.activity_main);
        recyclerView = findViewById(R.id.rcl_contacts);
        initData();
        adapter = new ContactsAdapter(this, list);
        LinearLayoutManager layout = new LinearLayoutManager(this);
        //设置布局管理器
        recyclerView.setLayoutManager(layout);
        //设置adapter
        recyclerView.setAdapter(adapter);
        //设置Item增加、移除动画
        recyclerView.setItemAnimator(new DefaultItemAnimator());
        //添加分割线
        recyclerView.addItemDecoration(new DividerItemDecoration(
                this, DividerItemDecoration.VERTICAL));
    }
    private void initData() {
        ContactsBean one = new ContactsBean(R.mipmap.ic_launcher, "商人", "计算星星数量忙得不可开交");
        list.add(one);
        ContactsBean two = new ContactsBean(R.mipmap.ic_launcher, "点灯人", "每分钟点一次灯熄一次灯");
        list.add(two);
        ContactsBean three = new ContactsBean(R.mipmap.ic_launcher, "地理学家", "只记录永恒的东西");
        list.add(three);
    }
}
```

运行程序,可以看到效果如图2.61所示。

图2.61　RecyclerView 示例

通过对比 ListView 的实现过程,可以看到 RecyclerView 更加简洁、灵活,代码也更加清晰,获取组件对象和赋值操作被放在两个方法中,有效地实现了解耦。直接使用 ViewHolder,减少了在程序中判断 findViewById 是否重复执行的步骤,代码也更加简单。

# 第3章 Android 事件处理

当用户进行触屏操作、单击按钮、在文本框中输入文本、在下拉式列表中选择一个条目、单击菜单项等操作时,都会发生界面事件。程序通过对界面事件做出反应来实现特定的任务,如切换界面、提交数据、根据用户选择做出响应等,这就是界面事件处理。Android 主要提供两种机制的事件处理方式:基于监听器的事件处理和基于回调的事件处理。同时,由于 Android 从线程安全的角度规定,只允许 UI 线程修改 Activity 主线程中的 UI 组件,当需要在新的线程中更新 Activity 主线程中的 UI 组件时,Android 提供了 Handler 回调机制来实现这一目标,在网络数据读取、游戏开发中 Handler 具有重要的作用。

## 3.1 基于监听器的事件处理

### 3.1.1 基于监听器的事件处理机制

Android 基于监听器的事件处理方式与 Java 类似,处理过程如图 3.1 所示。

☞ 事件源

能够产生事件的对象被称为事件源。在 Android 中,各种组件、Activity 本身都可以作为事件源。

☞ 事件

事件源会产生特定的事件,例如点击事件、长按事件、触屏事件等。一个事件源也可能产生多种事件,如在按钮上可以产生点击事件、长按事件、触屏事件等。

☞ 监视器

监视器用来对事件源进行监听,以便对发生的事件做出处理。事件源发生事件以后,监视器代替事件源对发生的事件进行处理。实现了对应监视器接口的类就能成为监视器,当事件源发生事件时,监视器会自动调用被实现的接口方法,对事件运行相应的代码就可以放置在该方法中。

在事件处理中,Android 实现接口的方式也与 Java 相同,可以通过外部类、内部类、匿名内部类等方式实现。

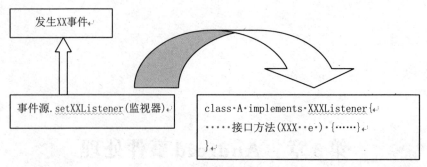

图 3.1 基于监听器的事件处理

### 3.1.2 onClick 点击事件处理、Toast

事件源：按钮、文本标签、编辑框等多种组件。

事件：单击事件。

监视器接口：OnClickListener。

接口方法：public void onClick(View view)。View 是发生事件的事件源对象，当多个事件源都需要响应 onClick 事件时，可以通过 view 判断是哪个事件源。

[例 3.1] 在 Android Studio 中新建一个项目，项目名为 OnClickDemo，在 res/layout/activity_main.xml 中设计布局，并在 MainActivity 中实现代码如下：

（1）activity_main.xml 代码清单。

```xml
<?xml version="1.0" encoding="utf-8"?>
<android.support.constraint.ConstraintLayout xmlns:android="http://schemas.android.com/apk/res/android"
    xmlns:app="http://schemas.android.com/apk/res-auto"
    xmlns:tools="http://schemas.android.com/tools"
    android:layout_width="match_parent"
    android:layout_height="match_parent"
    tools:context=".MainActivity">
    <Button
        android:id="@+id/btn_submit"
        android:layout_width="wrap_content"
        android:layout_height="wrap_content"
        android:layout_marginStart="8dp"
        android:layout_marginLeft="8dp"
        android:layout_marginTop="16dp"
        android:layout_marginEnd="264dp"
        android:layout_marginRight="264dp"
        android:text="提交"
        android:background="@color/colorPrimary"
```

```xml
        app:layout_constraintEnd_toEndOf="parent"
        app:layout_constraintHorizontal_bias="1.0"
        app:layout_constraintRight_toRightOf="parent"
        app:layout_constraintStart_toStartOf="parent"
        app:layout_constraintTop_toTopOf="parent"/>
    <TextView
        android:id="@+id/tv_help"
        android:layout_width="wrap_content"
        android:layout_height="30dp"
        android:layout_marginStart="8dp"
        android:layout_marginLeft="8dp"
        android:layout_marginTop="8dp"
        android:layout_marginEnd="8dp"
        android:layout_marginRight="8dp"
        android:text="获取帮助信息"
        android:textColor="@color/colorPrimary"
        app:layout_constraintEnd_toEndOf="parent"
        app:layout_constraintStart_toStartOf="parent"
        app:layout_constraintTop_toBottomOf="@+id/btn_submit"/>
</android.support.constraint.ConstraintLayout>
```

（2）MainActivity.java代码清单。

```java
package cc.turbosnail.onclickdemo;
import android.support.v7.app.AppCompatActivity;
import android.os.Bundle;
import android.view.View;
import android.widget.Button;
import android.widget.TextView;
import android.widget.Toast;
public class MainActivity extends AppCompatActivity {
    private Button btnSubmit;
    private TextView tvHelp;
    @Override
    protected void onCreate(Bundle savedInstanceState) {
        super.onCreate(savedInstanceState);
        setContentView(R.layout.activity_main);
        btnSubmit =findViewById(R.id.btn_submit);
        tvHelp = findViewById(R.id.tv_help);
```

```
btnSubmit.setOnClickListener(new View.OnClickListener() {
    @Override
    public void onClick(View v) {
        Toast.makeText(MainActivity.this, "你点击了Button", Toast.LENGTH_LONG).show();
    }
});
tvHelp.setOnClickListener(new View.OnClickListener() {
    @Override
    public void onClick(View v) {
        Toast.makeText(MainActivity.this,"你点击了 TextView", Toast.LENGTH_LONG).show();
    }
});
    }
}
```

运行程序,当点击按钮和文本标签时,会分别弹出不同的提示信息,如图3.2所示。

图3.2 点击事件的运行效果图

本例中,通过匿名内部类的方式分别为按钮 btnSubmit 和文本标签 tvHelp 添加了监视器,当OnClick点击事件发生在对应组件上时,接口方法就会被调用执行。本例中,对事件的响应是以 Toast 的形式弹出一个提示信息,在应用开发过程中,对事件的响应就是各种业务逻辑功能的实现。

Toast是Android中的一个常用组件,它用来在屏幕上弹出一个提示信息,显示一定时间后自动消失,被广泛地应用于对用户进行事件提醒。Toast的使用包括两个步骤:

(1)通过makeText 方法构建 Toast 对象。

Toast makeText(Context context, CharSequence text, int duration), context 为当前的上下

文环境，如当前的 Activity，text 为需要显示的文本字符串，duration 为显示时长，常用两个常量 Toast.LENGTH_SHORT 和 Toast.LENGTH_LONG。例如：

Toast toast = Toast.makeText(MainActivity.this,"提示",Toast.LENGTH_SHORT);

（2）通过 show 方法显示 Toast 信息。

toast.show();

### 3.1.3 onLongClick 长按事件处理

事件源：按钮、文本标签、编辑框等组件。

事件：长按事件。

监视器接口：OnLongClickListener。

接口方法：public boolean onLongClick(View v)。此方法返回 true，则表示事件响应结束，不会向下传递；返回 false，则表示事件响应向下传递，可以继续响应单击事件，即既响应长按事件，又响应单击事件。

[例3.2] 在 Android Studio 中新建一个项目，项目名为 OnLongClickDemo，在布局中放置一个按钮，id 为 "btn_test"，并在 MainActivity 中实现代码如下：

```java
package cc.turbosnail.onlongclickdemo;
import android.support.v7.app.AppCompatActivity;
import android.os.Bundle;
import android.view.Gravity;
import android.view.View;
import android.widget.Button;
import android.widget.Toast;
public class MainActivity extends AppCompatActivity {
    private Button button;
    @Override
    protected void onCreate(Bundle savedInstanceState) {
        super.onCreate(savedInstanceState);
        setContentView(R.layout.activity_main);
        button = findViewById(R.id.btn_test);
        button.setOnClickListener(new View.OnClickListener() {
            @Override
            public void onClick(View v) {
                Toast toast = Toast.makeText(MainActivity.this,"你单击了按钮",Toast.LENGTH_LONG);
                toast.setGravity(Gravity.CENTER, 0, 0);//调整 Toast 的显示位置
                toast.show();
            }
        });
```

```
        button.setOnLongClickListener(new View.OnLongClickListener() {
            @Override
            public boolean onLongClick(View v) {
                Toast.makeText(MainActivity.this, "你长按了按钮", Toast.LENGTH_LONG).show();
                return false;
            }
        });
    }
}
```

本例中为按钮 button 分别添加了点击事件和长按事件的两个监听器,运行程序,长按按钮可以看到 onLongClick 方法被执行,显示"你长按了按钮",在该显示结束后,还可以看到单击事件 onClick 方法也被执行,屏幕上还显示出"你单击了按钮"。运行效果如图 3.3 所示。

图 3.3　长按按钮事件运行效果

此时,如果将 onLongClick 方法的返回值修改为"return true",再次运行程序,长按按钮则可以看到 onLongClick 方法执行后,单击事件的 onClick 方法不会被执行。

### 3.1.4　onTouch 触屏事件处理

事件源:按钮、文本标签、编辑框等组件。

事件:触屏事件。

监视器接口:OnTouchListener。

接口方法:public boolean onTouch(View v, MotionEvent event)。此方法返回 true,则表示事件响应结束,不会向下传递;返回 false,则表示事件响应向下传递,可以继续响应长按事件、单击事件,即既响应触屏事件,又可以继续响应长按事件、单击事件。这三个事件的响应顺序为 onTouch->onLongClick->onClick。

MotionEvent参数是与用户触屏相关的事件,常见的动作常量如表3.1所示。

表3.1 MotionEvent动作常量

| 动作常量 | 描述 |
| --- | --- |
| MotionEvent.ACTION_DOWN | 第一个触点按下之后触发事件 |
| MotionEvent.ACTION_MOVE | 触点在屏幕上移动时触发事件 |
| MotionEvent.ACTION_UP | 当触点松开时触发事件 |
| MotionEvent.ACTION_POINTER_DOWN | 当屏幕上已经有触点处于按下的状态的时候,再有新的触点被按下时触发事件 |
| MotionEvent.ACTION_POINTER_UP | 当屏幕上有多个点被按住,松开其中一个点时(即非最后一个点被放开时)触发事件 |
| MotionEvent.ACTION_OUTSIDE | 用户触碰超出了正常的UI边界时触发事件 |

MotionEvent常用的方法如表3.2所示。

表3.2 MotionEvent常用方法

| 方法名 | 方法描述 |
| --- | --- |
| getAction() | 返回int型动作类型,其结果对应MotionEvent.ACTION_DOWN等动作常量 |
| getX()/getY() | 返回float型坐标位置,单位为像素,该位置(0,0)坐标为当前View的左上角,即返回的X和Y坐标是相对于当前View自身 |
| getRawX()/getRawY() | 返回float型坐标位置,单位为像素,该位置(0,0)坐标为屏幕的左上角,即返回的X和Y坐标是相对于当前屏幕的绝对坐标 |
| getSize() | 返回float型的接触面积大小,取值范围为0到1之间 |

**[例3.3]** 在Android Studio中新建一个项目,项目名为OnTouchDemo,在布局中放置一个文本标签,id为"tv_touch",并在MainActivity中实现代码如下:

(1) activity_main.xml布局代码。

```xml
<?xml version="1.0" encoding="utf-8"?>
<android.support.constraint.ConstraintLayout xmlns:android="http://schemas.android.com/apk/res/android"
    xmlns:app="http://schemas.android.com/apk/res-auto"
    xmlns:tools="http://schemas.android.com/tools"
    android:layout_width="match_parent"
    android:layout_height="match_parent"
    tools:context=".MainActivity">
    <TextView
        android:id="@+id/tv_touch"
        android:layout_width="match_parent"
        android:layout_height="300dp"
        android:background="#d3d3d3"
        android:text="onTouch测试"
```

```xml
            android:textAlignment="center"
            android:textColor="#0000FF"
            android:textSize="30dp"
            app:layout_constraintBottom_toBottomOf="parent"
            app:layout_constraintHorizontal_bias="1.0"
            app:layout_constraintLeft_toLeftOf="parent"
            app:layout_constraintRight_toRightOf="parent"
            app:layout_constraintTop_toTopOf="parent"
            app:layout_constraintVertical_bias="0.0"/>
</android.support.constraint.ConstraintLayout>
```

（2）MainActivity.java代码。

```java
package cc.turbosnail.ontouchdemo;
import android.support.v7.app.AppCompatActivity;
import android.os.Bundle;
import android.view.MotionEvent;
import android.view.View;
import android.widget.TextView;
import android.widget.Toast;
public class MainActivity extends AppCompatActivity {
    TextView tvTouch;
    @Override
    protected void onCreate(Bundle savedInstanceState) {
        super.onCreate(savedInstanceState);
        setContentView(R.layout.activity_main);
        tvTouch = findViewById(R.id.tv_touch);
        tvTouch.setOnTouchListener(new View.OnTouchListener() {
            @Override
            public boolean onTouch(View v, MotionEvent event) {
                if (event.getAction() == MotionEvent.ACTION_DOWN) {
                    Toast. makeText(MainActivity. this, " 触 屏 动 作 : 按 下 ", Toast. LENGTH_SHORT).show();
                }
                if (event.getAction() == MotionEvent.ACTION_UP) {
                    Toast. makeText(MainActivity. this, " 触 屏 动 作 : 松 开 ", Toast. LENGTH_SHORT).show();
                }
                if (event.getAction() == MotionEvent.ACTION_MOVE) {
```

```
                        Toast. makeText(MainActivity. this, " 触 屏 动 作：滑 动 ", Toast.
LENGTH_SHORT).show();
                        tvTouch.setText("当前坐标:" + event.getX() + "," + event.getY());
                    }
                    return true;
                }
            });
        }
    }
```

运行程序,在文本标签 tvTouch 上进行触屏操作,可以看到程序运行效果如图3.4所示。

图3.4 触屏事件运行效果

说明:setOnTouchListener 单独使用时,onTouch 方法需要返回true,才能保证"按下""松开""滑动"等动作都能响应,如果返回 false,则只会响应"按下"这一个动作。

当 setOnTouchListener 和 setOnClickListener 同时使用时,onTouch 的返回值要设为false,这样触屏动作可以全部响应,并且点击事件也会响应。

### 3.1.5 onFocusChange焦点变化事件处理

事件源:编辑框。

事件:焦点变化。

监视器接口:onFocuChangeListener。

接口方法:public void onFocusChange(View view, boolean b)。view 是触发该事件的事件源,b 用来判断得到焦点(true)和失去焦点(false)。

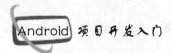

**[例3.4]** 在 Android Studio 中新建一个项目,项目名为 OnFocusChangeDemo,在布局中放置一个编辑框和一个文本标签,并在 MainActivity 中实现代码。

(1) activity_main.xml 布局代码。

```xml
<?xml version="1.0" encoding="utf-8"?>
<android.support.constraint.ConstraintLayout xmlns:android="http://schemas.android.com/apk/res/android"
    xmlns:app="http://schemas.android.com/apk/res-auto"
    xmlns:tools="http://schemas.android.com/tools"
    android:layout_width="match_parent"
    android:layout_height="match_parent"
    tools:context=".MainActivity">
    <EditText
        android:id="@+id/edt_input"
        android:layout_width="0dp"
        android:layout_height="wrap_content"
        android:text="EditText"
        app:layout_constraintLeft_toLeftOf="parent"
        app:layout_constraintRight_toRightOf="parent"
        app:layout_constraintTop_toTopOf="parent"/>
    <TextView
        android:id="@+id/tv_info"
        android:layout_width="wrap_content"
        android:layout_height="wrap_content"
        android:text="提示信息"
        app:layout_constraintLeft_toLeftOf="parent"
        app:layout_constraintRight_toRightOf="parent"
        app:layout_constraintTop_toBottomOf="@+id/edt_input"/>
</android.support.constraint.ConstraintLayout>
```

(2) MainActivity.java 代码。

```java
package cc.turbosnail.onfocuschangedemo;
import android.support.v7.app.AppCompatActivity;
import android.os.Bundle;
import android.view.View;
import android.widget.EditText;
import android.widget.TextView;
public class MainActivity extends AppCompatActivity {
    private EditText edtInput;
```

```java
private TextView tvInfo;
@Override
protected void onCreate(Bundle savedInstanceState) {
    super.onCreate(savedInstanceState);
    setContentView(R.layout.activity_main);
    edtInput = findViewById(R.id.edt_input);
    tvInfo = findViewById(R.id.tv_info);
    /*以下两行代码设置tvInfo可聚焦,当tvInfo获得焦点时,edtInput就会失去焦点*/
    tvInfo.setFocusable(true);//设置tvInfo可聚焦
    tvInfo.setFocusableInTouchMode(true);//设置tvInfo可触摸聚焦
    edtInput.setOnFocusChangeListener(new View.OnFocusChangeListener() {
        @Override
        public void onFocusChange(View view, boolean b) {
            if (b) {
                tvInfo.setText("编辑框EditText获得了焦点");
            } else {
                tvInfo.setText("编辑框EditText失去了焦点");
            }
        }
    });
    tvInfo.setOnClickListener(new View.OnClickListener() {
        @Override
        public void onClick(View view) {
            edtInput.requestFocus();//让edtInput获得焦点
        }
    });
}
```

运行程序,可以看到初始时编辑框拥有焦点,点击文本标签时,由于焦点转移到文本标签上,此时编辑框触发了失去焦点事件,当再次点击编辑框时,会触发得到焦点事件。在本例中,文本标签还通过onClick事件使文本框获得焦点。注意,设置Focusable属性后,第一次点击控件的时候,它将取得焦点,第二次点击的时候才响应点击事件。

程序运行效果如图3.5所示。

图3.5　编辑框焦点事件

### 3.1.6 onCreateContextMenu 弹出菜单事件处理

事件源：按钮、文本标签、编辑框等组件。

事件：上下文菜单生成。

监视器接口：onCreateContextMenuListener。

接口方法：public void onCreateContextMenu(ContextMenu menu, View v, ContextMenu.ContextMenuInfo menuInfo)。参数 menu 为准备创建的菜单，v 表示触发弹出菜单的组件，menuInfo 表示当前选中项 item 的信息。在这个方法中，可以添加 XML 格式定义的菜单，也可以添加 Java 代码直接生成的菜单。通过 Java 代码添加菜单项的方法如下：

menu.add(int groupId, int itemId, int order, CharSequence title)，参数 groupId 是组，可以将几个菜单项归为一组，以组的方式管理。参数 itemId 是菜单项编号，一个 itemId 对应一个菜单项，它是菜单项的唯一标示。参数 orderId 是菜单项的显示顺序，默认是 0，表示菜单项的显示按照 add 的添加顺序。参数 title 是菜单项中显示的文字标题。

【例 3.5】为组件弹出上下文菜单。在 Android Studio 中新建一个项目，项目名为 OnCreateContextMenuDemo，在布局中放置一个文本标签，text 为"长按弹出上下文菜单"，id 为"tv_context"，并在 MainActivity 中实现代码如下：

```java
package cc.turbosnail.oncreatecontextmenudemo;
import android.graphics.Color;
import android.support.v7.app.AppCompatActivity;
import android.os.Bundle;
import android.view.ContextMenu;
import android.view.MenuItem;
import android.view.View;
import android.widget.TextView;
import android.widget.Toast;
public class MainActivity extends AppCompatActivity {
    private TextView tvContext;
    @Override
    protected void onCreate(Bundle savedInstanceState) {
        super.onCreate(savedInstanceState);
        setContentView(R.layout.activity_main);
        tvContext = findViewById(R.id.tv_context);
        tvContext.setOnCreateContextMenuListener(new View.OnCreateContextMenuListener() {
            @Override
            public void onCreateContextMenu(ContextMenu menu, View v, ContextMenu.ContextMenuInfo menuInfo) {
                menu.add(0, 0, 0, "红色");
                menu.add(0, 1, 0, "绿色");
```

```
                menu.add(0, 2, 0, "蓝色");
            }
        });
    }
    @Override
    public boolean onContextItemSelected(MenuItem item) {
        switch (item.getItemId()) {
            case 0:
                tvContext.setBackgroundColor(Color.RED);
                Toast.makeText(MainActivity.this, "设置红色背景", Toast.LENGTH_SHORT).show();
                break;
            case 1:
                tvContext.setBackgroundColor(Color.GREEN);
                break;
            case 2:
                tvContext.setBackgroundColor(Color.BLUE);
                break;
            default:
                return super.onContextItemSelected(item);
        }
        return true;
    }
}
```

运行程序，在文本框上长按时会弹出上下文菜单，当用户选择菜单项时，可以根据 onContextItemSelected 方法的实现进行相应修改，本例中修改了文本标签的背景颜色，如图 3.6 所示。

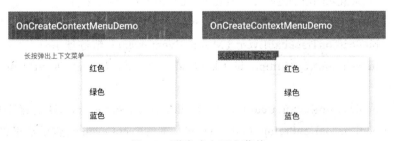

图 3.6 弹出式上下文菜单

在本例的 onCreateContextMenu 方法中为 menu 添加菜单项时，也可单独监听每个菜单项的 onMenuItemClick 事件实现事件响应，其接口为 MenuItem.OnMenuItemClickListener，添加监视器的方法为 setOnMenuItemClickListener，示例如下：

```
public void onCreateContextMenu(ContextMenu menu, View v, ContextMenu.ContextMenuInfo menuInfo) {
    MenuItem redItem=menu.add(0, 0, 0, "红色");
    redItem.setOnMenuItemClickListener(new MenuItem.OnMenuItemClickListener() {
        @Override
        public boolean onMenuItemClick(MenuItem item) {
            tvContext.setBackgroundColor(Color.RED);
            Toast.makeText(MainActivity.this, "设置红色背景", Toast.LENGTH_SHORT).show();
            return false;
        }
    });
}
```

## 3.2 基于回调的事件处理、LogCat

Android基于回调的事件处理是指当某个事件发生时,当前事件源自身的特定方法将被自动调用,以对事件进行响应。在这一过程中,事件源自己监听自己的事件,开发者不用再自定义监视器,只需要在对应的回调方法中实现响应代码即可。

基于回调的事件处理优先级一般低于基于监视器的方式,所以如果在基于监视器的监听中接口方法返回true,则对应的基于回调的方法将不会再执行。

在Android Studio中,可以通过菜单栏中的"Code"→"Override Methods"查看和添加组件的回调方法。

事件源:UI组件、Activity。

常用回调方法:

✓ public boolean onKeyDown(int keyCode, KeyEvent event),按下事件。
✓ public boolean onKeyUp(int keyCode, KeyEvent event),松开事件。
✓ public boolean onTouchEvent(MotionEvent event),触屏事件。
✓ public boolean onKeyLongPress(int keyCode, KeyEvent event),长按事件。
✓ public boolean onTrackballEvent(MotionEvent event),轨迹球事件。
✓ public boolean onKeyMultiple(int keyCode, int repeatCount, KeyEvent event),多次重复按键事件。
✓ public boolean onKeyShortcut(int keyCode, KeyEvent event),快捷键事件。
✓ public boolean onKeyPreIme(int keyCode, KeyEvent event),软键盘事件。

参数KeyEvent用来封装按键和按钮的事件,方法getKeyCode()可以获取当前按键的编码值,按键编码有数字、字母、功能键常量,如 KEYCODE_0、KEYCODE_A、KEYCODE_HOME等。方法getAction()可以获得当前动作的编码,动作编码常量有ACTION_DOWN、ACTION_MULTIPLE、ACTION_UP等。

【例3.6】在Activity和一个自定义按钮上响应回调事件。在Android Studio中新建一个项目,项目名为CallbackEventDemo。

(1) 在项目目录Java项目包名上点击右键,新建一个Java Class,类名为MyButton。自定义一个Button类,目录结构如图3.7所示。

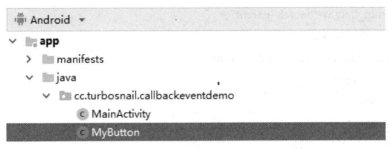

图3.7　自定义Button子类

MyButton.java代码如下:

```
package cc.turbosnail.callbackeventdemo;
import android.content.Context;
import android.support.v7.widget.AppCompatButton;
import android.util.AttributeSet;
import android.util.Log;
import android.view.MotionEvent;
import android.widget.Toast;
/*自定义Button时,编辑器提示将Button修改为AppCompatButton, AppCompatButton继承自Button,增加了动态背景等功能*/
public class MyButton extends AppCompatButton {
    public MyButton(Context context, AttributeSet attrs) {
        super(context, attrs);
    }
    public boolean onTouchEvent(MotionEvent event) {
        Log.i("info******", "MyButton的回调方法被调用");
        Toast.makeText(getContext(), "MyButton的回调方法被调用", Toast.LENGTH_LONG).show();
        return false;
    }
}
```

(2) activity_main.xml布局代码如下,在布局中添加了自定义的按钮组件,其写法是引用完整的组件路径和自定义类名"cc.turbosnail.callbackeventdemo.MyButton"。

```
<?xml version="1.0" encoding="utf-8"?>
<android.support.constraint.ConstraintLayout xmlns:android="http://schemas.android.com/apk/
```

```xml
    res/android"
    xmlns:app="http://schemas.android.com/apk/res-auto"
    xmlns:tools="http://schemas.android.com/tools"
    android:layout_width="match_parent"
    android:layout_height="match_parent"
    tools:context=".MainActivity">
    <cc.turbosnail.callbackeventdemo.MyButton
        android:id="@+id/btn_my"
        android:layout_width="wrap_content"
        android:layout_height="wrap_content"
        android:text="MyButton"
        app:layout_constraintBottom_toBottomOf="parent"
        app:layout_constraintLeft_toLeftOf="parent"
        app:layout_constraintRight_toRightOf="parent"
        app:layout_constraintTop_toTopOf="parent"/>
</android.support.constraint.ConstraintLayout>
```

（3）MainActivity代码如下，为自定义按钮添加了一个基于监视器的监听，此时由于它自身的回调方法也被实现，因此这个自定义按钮有两个监听。同时，为Activity本身实现了一个回调方法。

```java
package cc.turbosnail.callbackeventdemo;
import android.support.v7.app.AppCompatActivity;
import android.os.Bundle;
import android.util.Log;
import android.view.MotionEvent;
import android.view.View;
import android.widget.Button;
import android.widget.Toast;
public class MainActivity extends AppCompatActivity {
    private Button btn_my;
    @Override
    protected void onCreate(Bundle savedInstanceState) {
        super.onCreate(savedInstanceState);
        setContentView(R.layout.activity_main);
        btn_my = findViewById(R.id.btn_my);
        btn_my.setOnTouchListener(new View.OnTouchListener() {
            @Override
            public boolean onTouch(View v, MotionEvent event) {
```

```
            Log.i("info******", "MyButton的监听器方法被调用");
            Toast. makeText(MainActivity. this, "MyButton的监听器方法被调用", Toast.
LENGTH_LONG).show();
            return false;
        }
    });
}
/*重写MainActivity的回调事件*/
public boolean onTouchEvent(MotionEvent event) {
    if (event.getAction() == MotionEvent.ACTION_DOWN) {
        Log.i("info******", "MainActivity的回调方法被调用");
        Toast. makeText(MainActivity. this, "MainActivity的回调方法被调用", Toast.
LENGTH_LONG).show();
    }
    return true;
}
}
```

运行程序,点击按钮,控制台依次输出,如图3.8所示。

```
cc. turbosnail. callbackeventdemo I/info******: MyButton的监听器方法被调用
cc. turbosnail. callbackeventdemo I/info******: MyButton的回调方法被调用
cc. turbosnail. callbackeventdemo I/info******: MainActivity的回调方法被调用
```

图3.8 日志输出

在Android Studio中,点击底部标签"Logcat"可以打开LogCat视图窗口,程序运行中产生的日志信息都会在这里打印出来,它为开发者调试程序提供了一个很好的机制,可以通过简单的下拉列表查看不同日志,如图3.9所示。

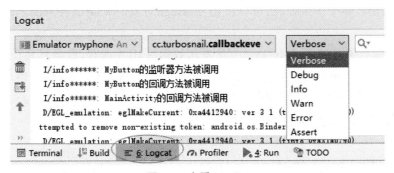

图3.9 查看LogCat

Android提供Log类来输出日志,日志从高到低可以分为6个级别,分别是VERBOSE、DEBUG、INFO、WARN、ERROR、ASSERT,对应的常用方法有5个。

☞ Log.v( ),黑色字体输出,任何一般消息都会输出。

☞ Log.d( ),蓝色字体输出,调试信息。

☞ Log.i( ),绿色字体输出,一般性的提示信息。

☞ Log.w( ),橙色字体输出,警告信息。

☞ Log.e( ),红色字体输出,错误信息。

以上方法可以在程序开发调试中使用,方法有两个参数,分别为标题和日志内容,用法示例如下:

---
Log.i("info******","当前值:"+s);//查看当前变量s的值

Log.v("verb******","方法已执行");

---

## 3.3 基于Handler的事件处理

Android只允许在UI主线程中修改UI组件。UI主线程是Activity启动时的线程,这意味着如果有子线程想要修改UI组件是不允许的,但是在实际开发中,经常需要根据子线程的运行情况修改主线程的UI组件内容,例如显示从网络下载的数据、定时更新动画效果等。因此,可以通过Android提供的Handler类的消息传递机制处理这样的问题。

### 3.3.1 Handler事件处理过程

Android的UI主线程应该主要用来负责处理界面显示、事件响应等操作,对于比较耗时的操作,例如网络数据下载、大量数据加载等,应该放在独立的子线程中进行,以避免UI线程被阻塞,甚至出现超时导致应用无响应的退出现象。这些子线程在执行后,一般都需要将结果传给主线程或者通知主线程其当前状态,然后由主线程更新界面显示的内容。

当需要在子线程中通知UI主线程修改界面组件时,Android提供基于Handler类的消息传递机制来实现,其实现过程为:

(1) 在UI主线程中创建Handler类的子类对象,重写handleMessage方法。该方法通过回调方式接收子线程传来的Message对象,即它一直处于监听状态,当子线程传来消息时该方法就会被调用。多个子线程时,可以根据Message对象的what属性判断当前是哪个子线程在发送消息。

(2) 在子线程中通过sendMessage方法发送数据给UI主线程,数据封装在Message对象中。Message对象的int型属性what一般用来作为标识值告诉UI主线程是哪一个子线程发送的数据,Object型的obj属性用来存储传递的数据。

Handler中的常用方法包括:

☞ public void handleMessage(Message msg)。

☞ public final boolean hasMessages(int what)。

☞ public final boolean sendMessage(Message msg)。

☞ public final boolean sendMessageDelayed(Message msg, long delayMillis)。

☞ public final boolean sendEmptyMessage(int what)。

☞ public final boolean sendEmptyMessageAtTime(int what, long uptimeMillis)。

☞ public final boolean sendEmptyMessageDelayed(int what, long delayMillis)。

**[例3.7]** 基于Handler实现一个简单的倒计时器。在Android Studio中新建一个项目，项目名为HandlerDemo。

（1）在activity_main.xml布局中放置一个文本标签，调整其大小、位置、字体，并设置初始显示值为"100"，表示倒计时初始时显示的值。

```xml
<?xml version="1.0" encoding="utf-8"?>
<android.support.constraint.ConstraintLayout xmlns:android="http://schemas.android.com/apk/res/android"
    xmlns:app="http://schemas.android.com/apk/res-auto"
    xmlns:tools="http://schemas.android.com/tools"
    android:layout_width="match_parent"
    android:layout_height="match_parent"
    tools:context=".MainActivity">
    <TextView
        android:id="@+id/tv_am"
        android:layout_width="140dp"
        android:layout_height="93dp"
        android:text="100"
        android:textSize="80sp"
        android:textAlignment="center"
        app:layout_constraintBottom_toBottomOf="parent"
        app:layout_constraintLeft_toLeftOf="parent"
        app:layout_constraintRight_toRightOf="parent"
        app:layout_constraintTop_toTopOf="parent"
        app:layout_constraintVertical_bias="0.098"/>
</android.support.constraint.ConstraintLayout>
```

（2）在MainActivity中创建Handler对象，等待接收消息并修改文本标签的显示内容，再创建一个计时器Timer对象，计时器线程启动后每隔1秒发送一个标识符what的值为1的消息，并把count的值封装在消息中一并发送出来。注意，Handler所在包为"android.os.Bundle"，导入包的语句是"import android.os.Bundle;"。

```java
package cc.turbosnail.handlerdemo;
import android.os.Bundle;
import android.os.Handler;
import android.os.Message;
import android.support.v7.app.AppCompatActivity;
import android.widget.TextView;
import java.util.Timer;
import java.util.TimerTask;
```

```java
public class MainActivity extends AppCompatActivity {
    private TextView textView;
    private int count = 100;//计数值
    private Handler handler = new Handler() {
        @Override
        public void handleMessage(Message msg) {
            switch (msg.what) {
                case 1:
                    textView.setText(String.valueOf((int) msg.obj));//修改UI组件
                    break;
                default:
                    break;
            }
        }
    };
    Timer timer = new Timer();
    TimerTask task = new TimerTask() {
        @Override
        public void run() {
            if (count > 0) {
                count--;
            } else {
                destroyTimer();
            }
            Message message = new Message();
            message.what = 1;//what只能放数字,一般用来做判断条件
            message.obj = count;//obj用来放对象,用来传值,可以是任何类型
            handler.sendMessage(message);
        }
    };
    @Override
    protected void onCreate(Bundle savedInstanceState) {
        super.onCreate(savedInstanceState);
        setContentView(R.layout.activity_main);
        textView = (TextView) findViewById(R.id.tv_am);
        timer.scheduleAtFixedRate(task, 1000, 1000);//启动子线程,开始计时
    }
    /*销毁计时器线程*/
    public void destroyTimer() {
```

```
        if (timer != null) {
            timer.cancel();
            timer = null;
        }
    }
}
```

运行程序,可以看到计时器开始倒计时,如图3.10所示。

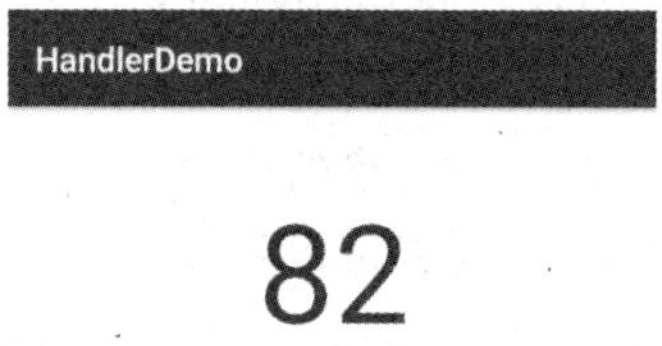

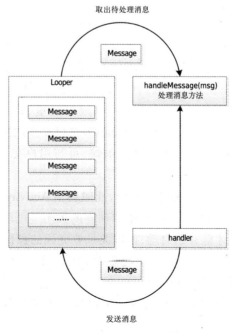

图3.10　倒计时运行效果

### 3.3.2　对Handler的进一步说明

Handler发送消息Message后,多个Message会进入消息队列MessageQueue中排队,MessageQueue采用先进先出的方式管理消息队列,但是它需要Looper来负责读取消息,再发送给该消息的Handler进行处理。在UI主线程中,系统默认创建了一个Looper对象,所以我们只需要通过Handler发送和接收数据即可,但是如果要在非UI主线程中使用Handler,就必须通过Android提供的Looper.prepare()创建一个Looper对象,再通过Looper.loop()运行消息循环才能实现取值的功能。创建MessageQueue对象的代码已经被封装在Looper的构造方法中,所以不用再单独创建MessageQueue对象了,如图3.11所示。

图3.11　Handler处理流程

## 3.4 界面跳转

### 3.4.1 Activity 的生命周期

Activity 是 Android 中用来负责与用户交互的基础组件,一个 Activity 可理解为一个覆盖整个屏幕的容器,它代表当前的一个屏幕,用户界面上的内容均可以放置在 Activity 中显示和进行事件响应。Activity 是 Android 的四大程序组件之一(Android 的四大程序组件分别是 Activity、Service、BroadCastReceiver、ContentProvider)。一款应用软件往往会有多屏,例如登录界面、欢迎界面、各类功能界面等,这其中的每一屏一般就是一个独立的 Activity。

Activity 从创建到销毁的过程是一个完整的生命周期,Android 定义了 Activity 生命周期各个阶段对应的方法供开发者调用,开发者可以把功能代码写在这些方法中,这些方法会在对应的生命周期阶段被自动调用,但是 Activity 本身的生命周期管理由 Android 系统负责,开发者并不能干预。如:在 Activity 启动时,会调用生命周期方法中的 onCreate()方法,我们可以把初始化和设置界面的代码放在该方法中,则当前 Activity 启动时,这些代码也会被执行。

```
protected void onCreate(Bundle savedInstanceState) {
    super.onCreate(savedInstanceState);
    setContentView(R.layout.activity_main);//启动时设置界面
}
```

Activity 完整的生命周期如图 3.12 所示。

在 Activity 的生命周期中,进入运行状态后,该 Activity 位于屏幕前端,获得焦点,用户可见;进入暂停状态后,依然可见,但是不能获得焦点;进入停止状态后,该 Activity 不可见,失去焦点;到了销毁状态,该 Activity 结束,被销毁。在这个过程中,对应的方法将被调用,这些方法包括:

(1) onCreate。Activity 创建时被调用,此时 Activity 处于不可见状态。在这里可以做一些初始化的操作,比如使用 setContentView 加载布局、对一些控件和变量进行初始化等。由生命周期图可以看出,onCreate 方法只在 Activity 创建时执行一次。

(2) onStart。Activity 正在启动时被调用,Activity 已处于可见状态,但还没有在前端,无法与用户进行交互。一些初始化的操作也可以放在这里,但习惯上初始化操作被放在 onCreate 方法里。由生命周期图可以看出,onStart 方法可以多次执行。

(3) onResume。Activity 已经可见且处于前端,可与用户交互。开启动画的操作、打开独占设备的操作可以放在这里。

(4) onPause。Activity 正在停止时被调用,此时当前 Activity 还在前端可见。一些简单的数据存储操作、程序状态的保存、独占设备和动画的关闭等可以放在这个阶段完成。本方法执行完以后别的 Activity 才能开始启动,所以此处要完成的操作不宜过多。

(5) onStop。Activity 停止或者完全被覆盖时调用。此时 Activity 不可见,仅在后台存在。这个阶段可以做一些资源回收的操作。

(6) onDestroy。Activity 被销毁时调用。可以进行资源释放和回收的操作,onDestroy

方法执行之后,Acivity对象被销毁,不再存在。

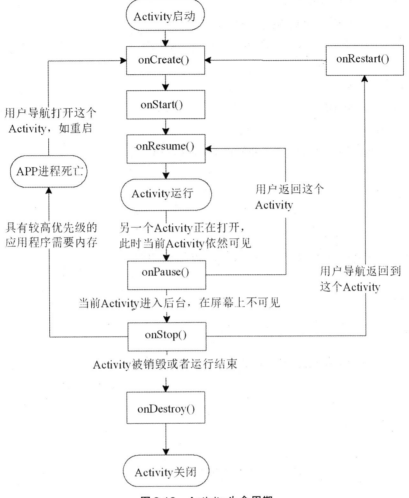

图3.12 Activity生命周期

（7）onRestart。Activity重新开始时调用,由不可见状态变为可见状态。从后一个Activity切换回前一个Activity或者用户按Home键切换到桌面后又切换回来时这个方法会被调用。

[例3.8] Activity生命周期示例。在Android Studio中新建一个项目,项目名为ActivityLifeCycleDemo。首先在MainActivity.java代码中重新设计各个生命周期方法,然后通过LogCat的形式输出提示信息。

```
package cc.turbosnail.activitylifecycledemo;
import android.support.v7.app.AppCompatActivity;
import android.os.Bundle;
import android.util.Log;
public class MainActivity extends AppCompatActivity {
    @Override
```

```java
protected void onCreate(Bundle savedInstanceState) {
    super.onCreate(savedInstanceState);
    setContentView(R.layout.activity_main);
    Log.i("ActivityLifeCycle***","onCreate方法被调用");
}
@Override
protected void onStart() {
    super.onStart();
    Log.i("ActivityLifeCycle***","onStart方法被调用");
}
@Override
protected void onResume() {
    super.onResume();
    Log.i("ActivityLifeCycle***","onResume方法被调用");
}
@Override
protected void onPause() {
    super.onPause();
    Log.i("ActivityLifeCycle***","onPause方法被调用");
}
@Override
protected void onStop() {
    super.onStop();
    Log.i("ActivityLifeCycle***","onStop方法被调用");
}
@Override
protected void onRestart() {
    super.onRestart();
    Log.i("ActivityLifeCycle***","onRestart方法被调用");
}
@Override
protected void onDestroy() {
    super.onDestroy();
    Log.i("ActivityLifeCycle***","onDestroy方法被调用");
}
}
```

运行程序,可以看到onCreate、OnStart、OnResume方法首先被调用执行;关闭程序,可以看到onPause、onStop、onDestroy方法被调用执行,如图3.13所示。

```
cc.turbosnail.activitylifecycledemo  I/ActivityLifeCycle***: onCreate方法被调用
cc.turbosnail.activitylifecycledemo  I/ActivityLifeCycle***: onStart方法被调用
cc.turbosnail.activitylifecycledemo  I/ActivityLifeCycle***: onResume方法被调用
cc.turbosnail.activitylifecycledemo  I/ActivityLifeCycle***: onPause方法被调用
cc.turbosnail.activitylifecycledemo  I/ActivityLifeCycle***: onStop方法被调用
cc.turbosnail.activitylifecycledemo  I/ActivityLifeCycle***: onDestroy方法被调用
```

图3.13　Activity生命周期运行示例

### 3.4.2　Activity跳转与传值跳转

当一个程序中有多个Activity时,这些Activity以栈的形式排列,如图3.14所示,并遵守以下原则:

☞ 处于栈顶端的Activity可见并可与用户交互。

☞ 启动一个新的Activity,它将进入栈顶。

☞ 先进后出原则。

☞ 用户通过Back按钮返回,或者栈顶的Activity结束,下一个Activity就会位于栈的首位并进入运行状态。

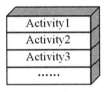

图3.14　Activity栈

程序从第一个Activity开始启动执行,在这个Activity中可以启动另一个Activity,通过这样的关联,多个Activity之间可以进行切换,使得程序可以完成多种任务。当从Activity1启动Activity2时,其执行过程如图3.15所示。

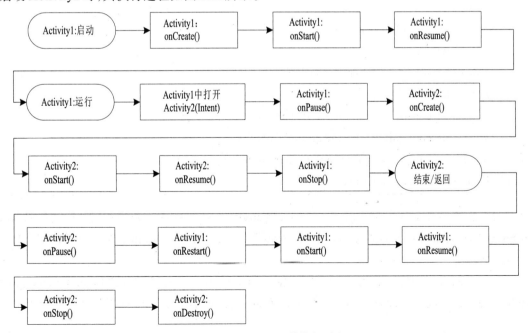

图3.15　Activity跳转执行过程

Android通过Intent对象来封装"意图",通过Intent对象统一组件之间的切换请求,同时,通过Bundle对象来封装数据,以Intent对象携带Bundle对象的形式实现数据交换。

**1. Activity中启动其他Activity进行跳转**

在当前Activity启动其他Activity的方法是startActivity(Intent intent),Intent对象用来表达从一个程序组件到另一个程序组件的意图,例如:

Intent intent = new Intent(MainActivity.this, SecondActivity.class);表示建立从MainActivity跳转到SecondActivity的意图,但是还没有执行跳转。

在创建了"意图"对象后,就可以通过跳转语句来进行跳转,例如:

startActivity(intent);从当前MainActivity跳转启动SecondActivity。

当启动一个新的Activity时,可以通过Activity的finish()方法通知系统销毁当前Activity,这样可以减少在队列中的Activity数量,其语句可以写为"finish();"。

[例3.9] Activity跳转示例。用户点击按钮,Activity跳转。

(1)在Android Studio中新建一个项目,项目名为IntentDemo。在布局中放置一个用来跳转的按钮,修改布局文件activity_main.xml代码如下:

```
<?xml version="1.0" encoding="utf-8"?>
<android.support.constraint.ConstraintLayout xmlns:android="http://schemas.android.com/apk/res/android"
    xmlns:app="http://schemas.android.com/apk/res-auto"
    xmlns:tools="http://schemas.android.com/tools"
    android:layout_width="match_parent"
    android:layout_height="match_parent"
    tools:context=".MainActivity">
    <Button
        android:id="@+id/btn_jump"
        android:layout_width="0dp"
        android:layout_height="wrap_content"
        android:text="跳转"
        app:layout_constraintLeft_toLeftOf="parent"
        app:layout_constraintRight_toRightOf="parent"
        app:layout_constraintTop_toTopOf="parent" />
</android.support.constraint.ConstraintLayout>
```

(2)新建第二个Activity,命名为SecondActivity。在项目目录列表的"java"→包名"cc.turbosnail.intentdemo"上点击右键,选择"New"→"Activity"→"Empty Activity",如图3.16所示。

在弹出的窗口中输入Activity的名称"SecondActivity",此时可以看到Layout Name自动修改为activity_second,即自动生成了对应的布局文件activity_second.xml,如图3.17所示。

第 3 章 Android 事件处理

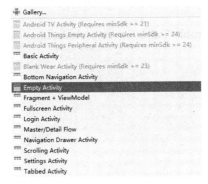

图 3.16 创建新的 Activity

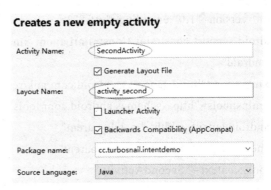

图 3.17 新 Activity 命名

创建完成后，项目目录中就有了 2 个 Activity 和 2 个对应的布局文件，SecondActivity 的代码结构和 MainActivity 类似，它设置了 activity_second 为当前显示视图。项目当前的目录结构如图 3.18 所示。

图 3.18 两个 Activity 目录结构

注意，Activity 作为 Android 的基础程序组件，每一个都需要在 AndroidManifest.xml 文件中注册，在 Android Studio 中创建 Activity 时，它自动生成布局文件的同时还自动进行了注册。打开 AndroidManifest.xml 文件可以看到，在 <application> 标签中添加了如下语句：
<activity android:name=".SecondActivity"></activity>。

由这个文件也可以看到，当有多个 Activity 被注册时，程序启动的 Activity 由以下代码决定：

```
<intent-filter>
    <action android:name="android.intent.action.MAIN"/>
    <category android:name="android.intent.category.LAUNCHER"/>
</intent-filter>
```

如果要修改启动时加载的第一个 Activity，只需要将以上代码放到对应的 <activity> 标签中即可，一个应用中只能有一个 Activity 作为启动的 Activity。

（3）可以根据需要修改 activity_second.xml 的代码，作为跳转的目标界面，设置布局如下：

```xml
<?xml version="1.0" encoding="utf-8"?>
<android.support.constraint.ConstraintLayout xmlns:android="http://schemas.android.com/apk/res/android"
    xmlns:app="http://schemas.android.com/apk/res-auto"
    xmlns:tools="http://schemas.android.com/tools"
    android:layout_width="match_parent"
    android:layout_height="match_parent"
    tools:context=".SecondActivity">
    <TextView
        android:layout_width="wrap_content"
        android:layout_height="wrap_content"
        android:text="欢迎来到第二个Activity!"
        app:layout_constraintEnd_toEndOf="parent"
        app:layout_constraintStart_toStartOf="parent"
        app:layout_constraintTop_toTopOf="parent"/>
</android.support.constraint.ConstraintLayout>
```

（4）在MainActivity.java中添加事件监听按钮，发生事件时进行跳转。

```java
package cc.turbosnail.intentdemo;
import android.content.Intent;
import android.support.v7.app.AppCompatActivity;
import android.os.Bundle;
import android.view.View;
import android.widget.Button;
public class MainActivity extends AppCompatActivity {
    Button btn_jump;
    @Override
    protected void onCreate(Bundle savedInstanceState) {
        super.onCreate(savedInstanceState);
        setContentView(R.layout.activity_main);
        btn_jump = findViewById(R.id.btn_jump);
        btn_jump.setOnClickListener(new View.OnClickListener() {
            @Override
            public void onClick(View v) {
                Intent intent = new Intent(MainActivity.this, SecondActivity.class);//创建意图
                startActivity(intent);//跳转
                finish();//通知销毁当前的MainActivity
            }
```

            });
        }
    }
}

（5）运行程序，点击按钮时，Activity 进行了跳转，效果如图 3.19 所示。

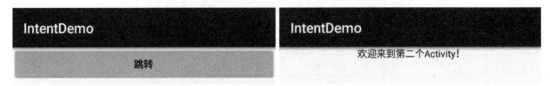

图 3.19　Activity 跳转运行效果

### 2. 基于 Bundle 在 Activity 跳转时传值

在应用开发中，经常需要在界面跳转时将数据传递给第二个 Activity，以便于进一步处理。Intent 可以携带 Bundle 类型的数据对象来进行数据传递，Bundle 以键值对（key，value）的形式存放数据信息。其实现步骤为：

（1）创建 Bundle 对象，根据数据类型存入数据。

Bundle bundle = new Bundle();
bundle.putString("name", "zhangsan");
bundle.putInt("age", 19);

（2）把封装好的 bundle 对象添加给 Intent，跳转时数据就会传递。

Intent intent = new Intent(MainActivity.this,SecondActivity.class);
intent.putExtras(bundle);
startActivity(intent);

（3）跳转后，在目标 Activity 中通过对应的方法接收数据。接收数据时，通过封装时（key，value）结构中的 key 来取值，注意封装数据和获取数据的 set/get 方法的数据类型应保持一致。

Intent intent = getIntent();//直接获取,不是再创建一个新的 Intent 对象
Bundle bundle = intent.getExtras();
String name = bundle.getString("name");
int age = bundle.getInt("age");

Bundle 支持多种数据类型的存和取操作，对常用数据类型，如 boolean、byte、char、double、float、int、long、String 等都封装了对应的 5 个方法，方法声明的示例如下：

☞ void putBoolean(String key, boolean value)

☞ boolean getBoolean(String key)

☞ boolean getBoolean(String key, boolean defaultValue)

☞ void putBooleanArray(String key, boolean[] value)

☞ boolean[] getBooleanArray(String key)

defaultValue 参数的作用是当没有取到值时，默认返回一个值，作为取值结果。

【例 3.10】Activity 基于 Bundle 传值示例。从一个 Activity 中把用户输入的姓名和年龄信息传递到跳转后的第二个 Activity 中，并显示出来。

（1）在 Android Studio 中新建一个项目，项目名为 IntentBundleDemo。在 activity_main.xml 布局中放置界面组件：两个文本标签用来提示"用户名"和"年龄"，两个编辑框用来接收用户输入，一个按钮用来进行跳转，界面设计如图 3.20 所示。

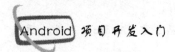

图 3.20　Bundle 传值界面设计

组件 id 属性声明如下：

&lt;EditText android:id="@+id/edt_name"/&gt;

&lt;EditText android:id="@+id/edt_age"/&gt;

&lt;Button android:id="@+id/btn_jump"/&gt;

（2）新建第二个 Activity，命名为 SecondActivity。在项目目录列表的"java"→包名"cc.turbosnail.intentbundledemo"上点击右键，选择"New"→"Activity"→"Empty Activity"。

在 SecondActivity 的布局文件 activity_second.xml 中放置一个文本标签，用来显示传值过来的用户名和年龄信息，其 id 属性设置如下：

&lt;TextView android:id="@+id/tv_info"/&gt;

（3）在 MainActivity 中响应事件，进行传值，代码清单如下：

```
package cc.turbosnail.intentbundledemo;
import android.content.Intent;
import android.support.v7.app.AppCompatActivity;
import android.os.Bundle;
import android.view.View;
import android.widget.Button;
import android.widget.EditText;
public class MainActivity extends AppCompatActivity {
    private EditText edtName;
    private EditText edtAge;
    private Button btnJump;
    @Override
    protected void onCreate(Bundle savedInstanceState) {
        super.onCreate(savedInstanceState);
        setContentView(R.layout.activity_main);
```

```java
        edtName = findViewById(R.id.edt_name);
        edtAge = findViewById(R.id.edt_age);
        btnJump = findViewById(R.id.btn_jump);
        btnJump.setOnClickListener(new View.OnClickListener() {
            @Override
            public void onClick(View v) {
                Bundle bundle = new Bundle();//创建传值的Bundle对象
                bundle.putString("name", edtName.getText().toString());
                bundle.putInt("age", Integer.parseInt(edtAge.getText().toString()));
                Intent intent = new Intent(MainActivity.this, SecondActivity.class);
                intent.putExtras(bundle);//把bundle添加给intent
                startActivity(intent);
            }
        });
    }
}
```

（4）在SecondActivity中获取Intent对象，取出传递过来的值，赋值给文本标签进行显示。代码清单如下：

```java
package cc.turbosnail.intentbundledemo;
import android.content.Intent;
import android.support.v7.app.AppCompatActivity;
import android.os.Bundle;
import android.widget.TextView;
public class SecondActivity extends AppCompatActivity {
    private TextView textView;
    @Override
    protected void onCreate(Bundle savedInstanceState) {
        super.onCreate(savedInstanceState);
        setContentView(R.layout.activity_second);
        textView = findViewById(R.id.tv_info);
        Intent intent = getIntent();//获取intent
        Bundle bundle = intent.getExtras();
        String name = bundle.getString("name");//取值
        int age = bundle.getInt("age");//取值
        textView.setText("姓名:" + name + " 年龄:" + age);
    }
}
```

程序运行效果如图3.21所示。

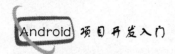

图3.21 Bundle传值运行效果

### 3. 基于Bundle返回数据到前一个Activity

当在一个Activity中启动一个新的Activity后，又需要把新Activity中的数据传回原Activity中时，可以通过startActivityForResult、onActivityResult、setResult三个方法组合来实现。

（1）startActivityForResult（Intent intent，int requestCode）。该方法在第一个Activity中执行。通过第一个参数intent进行跳转，第二个参数requestCode取大于或等于0的int型数值，这个值用来标识要启动的新Activity，当新Activity往回传值时，这个值用来确认原来的Activity。

（2）protected void onActivityResult（int requestCode，int resultCode，Intent data）。该方法在第一个Activity中重写。第一个参数requestCode用于与startActivityForResult中的requestCode值进行比较，以便确认数据是从哪个Activity发出的。第二个参数resultCode是由新Activity通过其setResult（）方法返回的，用于标识是哪一个activity返回的值。第三个参数用来接收Intent对象。

（3）setResult（int resultCode，Intent data）。该方法在新Activity中执行。通过resultCode和Intent对象实现跳转回原Activity的功能，并将数据通过data传回。resultCode一般可以用默认常量RESULT_OK，其值为-1。

【例3.11】基于Bundle返回数据到前一个Activity示例。从一个Activity中把用户输入的姓名和年龄信息传递到跳转后的第二个Activity中，并显示出来。

（1）在Android Studio中新建一个项目，项目名为IntentBundleReturnDemo。在activity_main.xml布局中放置界面组件：一个文本标签用来显示当前金额，一个按钮用来进行跳转，界面设计如图3.22所示。组件id属性声明如下：

```
<TextView android:id="@+id/tv_info"/>
<Button android:id="@+id/bt_jump"/>
```

（2）创建SecondActivity。在项目目录列表的"java"→包名"cc.turbosnail.intentbundlereturndemo"上点击右键，选择"New"→"Activity"→"Empty Activity"。

在SecondActivity的布局文件activity_second.xml中放置一个编辑框用来让用户输入金额、1个按钮用来返回前一个Activity，如图3.23所示。组件id属性声明如下：

```xml
<EditText android:id="@+id/edt_input"/>
<Button android:id="@+id/bt_back"/>
```

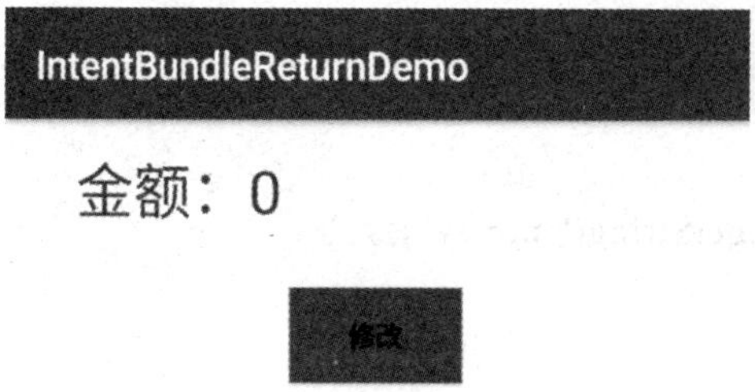

图 3.22　activity_main 界面设计 　　　图 3.23　activity_second 界面设计

（3）在 MainActivity 中响应事件，跳转到第二个 Activity，并重写 onActivityResult 方法，等待处理返回数据。代码清单如下：

```java
package cc.turbosnail.intentbundlereturndemo;
import android.content.Intent;
import android.support.v7.app.AppCompatActivity;
import android.os.Bundle;
import android.view.View;
import android.widget.Button;
import android.widget.TextView;
public class MainActivity extends AppCompatActivity {
    private Button button;
    private TextView textView;
    @Override
    protected void onCreate(Bundle savedInstanceState) {
        super.onCreate(savedInstanceState);
        setContentView(R.layout.activity_main);
        textView = findViewById(R.id.tv_info);
        button = findViewById(R.id.bt_jump);
        button.setOnClickListener(new View.OnClickListener() {
            @Override
            public void onClick(View v) {
                Intent intent = new Intent(MainActivity.this, SecondActivity.class);
                startActivityForResult(intent, 1);//1用来作为标识符
            }
        });
    }
    @Override
    protected void onActivityResult(int requestCode, int resultCode, Intent data) {
```

```java
        super.onActivityResult(requestCode, resultCode, data);
        if (requestCode == 1 && resultCode == RESULT_OK) { //使用标识符 1 和 RESULT_OK
            Bundle bundle = data.getExtras();
            if (bundle != null) {
                textView.setText("金额:" + bundle.getString("money"));
            }
        }
    }
}
```

（4）在 SecondActivity 中通过 setResult 方法回传数据。代码清单如下：

```java
package cc.turbosnail.intentbundlereturndemo;
import android.content.Intent;
import android.support.v7.app.AppCompatActivity;
import android.os.Bundle;
import android.view.View;
import android.widget.Button;
import android.widget.EditText;
public class SecondActivity extends AppCompatActivity {
    private EditText editText;
    private Button button;
    @Override
    protected void onCreate(Bundle savedInstanceState) {
        super.onCreate(savedInstanceState);
        setContentView(R.layout.activity_second);
        editText = findViewById(R.id.edt_input);
        button = findViewById(R.id.bt_back);
        button.setOnClickListener(new View.OnClickListener() {
            @Override
            public void onClick(View v) {
                Intent intent = getIntent();//获取 intent 对象
                Bundle bundle = new Bundle();
                bundle.putString("money", editText.getText().toString());
                intent.putExtras(bundle);
                setResult(RESULT_OK, intent);//回传数据
                finish();
            }
```

```
    });
  }
}
```

运行程序,初始时显示金额为"0",点击"修改"按钮,跳转到第二个Activity中,在文本框中输入数字金额,点击返回第一个Activity,可以看到此时输入值被传回,并显示在文本标签上,如图3.24所示。

图 3.24　基于Bundle返回数据运行效果

### 3.4.3　Intent打开系统应用:拨号、短信、相机

Intent可传递的数据包括动作(Action)、分类(Category)、数据(Data)、类型(Type)、组件(Component)、扩展信息(Extra)和标志(Flags)七种类型。除了显示指定跳转的目标组件之外,还可以不明确指定目标,而是通过设置动作、分类、数据的方式,让系统来筛选符合设置的目标组件进行隐式跳转。

可以在项目配置文件AndroidManifest.xml中为基础组件自定义Intent过滤器,设置Action和Category的相关属性,然后在跳转时由系统根据属性匹配自动跳转,例如:

```
<activity android:name=".TestActivity">
<intent-filter>
    <action android:name="cc.turbosnail.Test"/>
    <category android:name="android.intent.category.DEFAULT"/>
</intent-filter>
</activity>
```

此时,自定义Action的name属性,建议用"完整包名.Activity名"的结构作为属性值,如"cc.turbosnail.Test"。Category必须和Action配合使用,它的name属性一般可以使用默认值"android.intent.category.DEFAULT"。在从MainActivity向它跳转时,可以通过如下代码实现隐式跳转,让系统通过Action去匹配目标:

```
Intent intent = new Intent();
intent.setAction("cc.turbosnail.Test");
startActivity(intent);
```

Data属性允许通过协议名称由系统启动对应的Activity,它也属于隐式跳转。例如可以将Activity声明成支持一个自定义协议"my-protocol"的配置代码如下:

```xml
<activity android:name=".Test2Activity">
    <intent-filter>
        <action android:name="cc.turbosnail.Test2"/>
        <category android:name="android.intent.category.DEFAULT"/>
        <data android:scheme="my-protocol"/>
    </intent-filter>
</activity>
```

在跳转时,"my-protocol"作为协议名称,Uri数据"test"写在协议后面,通过setData方法进行设置,系统会自动匹配支持"my-protocol"的Activity来作为跳转目标。如果系统中有多个Activity都支持对应协议,即具有相同的data属性,则会弹出窗口让用户进行选择。

```java
Intent intent=new Intent();
Uri uri= Uri.parse("my-protocol:test");
intent.setData(uri);
startActivity(intent);
```

Android系统为开发者提供了丰富的Action和Data属性,可以方便地实现调用浏览器、拨打电话、发送短信、打开系统相机等功能,一些用法列表如下:

(1) Intent.ACTION_VIEW:根据传入的数据类型打开相应的Activity进行数据显示。

```java
Intent intent = new Intent();
intent.setAction(Intent.ACTION_VIEW);
intent.setData(Uri.parse("http://www.baidu.com"));//打开浏览器
//intent.setData(Uri.parse("geo:19,110"));//打开地图(北纬,东经)
startActivity(intent);
```

(2) Intent.ACTION_DIAL:打开电话拨号面板。

```java
Intent intent = new Intent();
intent.setAction(Intent.ACTION_DIAL);
intent.setData(Uri.parse("tel:13807601111"));//打开拨号面板
startActivity(intent);
```

(3) Intent.ACTION_SENDTO:打开短信发送面板。

```java
Intent intent = new Intent();
intent.setAction(Intent.ACTION_SENDTO);//打开短信面板
intent.setData(Uri.parse("smsto:13807601111"));
intent.putExtra("sms_body","短信内容");
startActivity(intent);
```

（4）Intent.ACTION_PICK：打开多媒体选择器。

```
Intent intent = new Intent();
intent.setAction(Intent.ACTION_PICK);
intent.setData(MediaStore.Images.Media.EXTERNAL_CONTENT_URI); // 打开图片库
//intent.setType("image/*");//选择照片
//intent.setType("audio/*");//选择音频
//intent.setType("video/*");//选择视频（MP4、3gp）
//intent.setType("video/;image/");//选择视频和图片
startActivityForResult(intent, 2);//用于返回选择的对象
```

（5）MediaStore.ACTION_IMAGE_CAPTURE：打开系统相机。

```
Intent intent = new Intent();
intent.setAction(MediaStore.ACTION_IMAGE_CAPTURE);//打开系统相机
startActivityForResult(intent, 1);//用于返回图片对象
```

**[例3.12]** 通过Intent打开系统手机拨号面板。在Android Studio中新建一个项目，名称为DialDemo。在activity_main.xml布局文件中放置一个按钮，然后在MainActivity代码中实现事件监听，当用户点击按钮时打开系统拨号面板。主要代码部分如下：

```java
public class MainActivity extends AppCompatActivity {
    Button button;
    @Override
    protected void onCreate(Bundle savedInstanceState) {
        super.onCreate(savedInstanceState);
        setContentView(R.layout.activity_main);
        button= findViewById(R.id.button);
        button.setOnClickListener(new View.OnClickListener() {
            @Override
            public void onClick(View v) {
                Intent intent = new Intent();
                intent.setAction(Intent.ACTION_DIAL);
                intent.setData(Uri.parse("tel:13807601111"));//打开拨号面板
                startActivity(intent);
            }
        });
    }
}
```

运行程序，可以看到系统拨号面板被打开，在模拟器上的显示如图3.25所示。

图 3.25　启动手机拨号面板

## 3.5　Activity四种启动模式

当通过startActivity等方法启动一个Activity时，默认情况下这个新启动的Activity将进入栈顶，处于可见并能与用户交互的状态，但是，如果这个Activity的实例对象有一个已经存在或者正处于栈中的某个位置，则可以根据业务逻辑需要，选择新创建实例对象或者直接使用栈中已经存在的实例对象。因此，Android为Activity提供了四种启动模式来声明Activity如何实例化。其设置方法为：在项目目录的AndroidManifest.xml配置文件中，为<Activity>标签添加"android：launchMode"属性，属性可取值为standard、singleTop、singleTask、singleInstance。示例如下：

　　<activity android:name=".MainActivity" android:launchMode="singleTask">

### 3.5.1　standard模式

standard是默认的启动模式，每次启动一个新的Activity时，它就会创建一个新的实例对象，并处于栈顶的位置。对于使用standard模式来说，系统不会在乎这个Activity是否已经有实例对象在栈中存在。例如：

```
public void onClick(View v) {
    Intent intent=new Intent(MainActivity.this, MainActivity.class);
    startActivity(intent);
}
```

MainActivity跳转时继续启动一个MainActivity的实例对象，此时点击按钮，可以看到屏幕有切换效果，连续点击三次按钮，则栈中一共出现四个MainActivity实例，最后创建的实例对象位于栈顶，栈结构如图3.26所示。此时，如果需要退出程序的话，则需要点击四次返回键。

| |
|---|
| MainActivity实例 |
| MainActivity实例 |
| MainActivity实例 |
| MainActivity实例 |

图3.26　standard启动模式示意图

创建的实例对象可以在Android Monitor控制台的Logcat中看到，如图3.27所示。

```
9260-9299/cc.turbosnail.onclickdemo D/OpenGLRenderer: endAllActiveAnimators on 0x9634d380 (RippleDrawable) with handle 0x96347870
9260-9299/cc.turbosnail.onclickdemo D/OpenGLRenderer: endAllActiveAnimators on 0x9535a700 (RippleDrawable) with handle 0x963473e0
9260-9299/cc.turbosnail.onclickdemo D/OpenGLRenderer: endAllActiveAnimators on 0x94da0700 (RippleDrawable) with handle 0x9537f070
```

图3.27　Logcat中查看实例对象

### 3.5.2　singleTop模式

singleTop模式下启动Activity时，如果这个Activity的实例对象已经处于栈顶的位置，则直接使用这个已经存在的实例对象，不再新建。否则，无论栈中是否还有该Activity的实例对象，都会新建一个进入栈顶位置。

以上例中的MainActivity为例，修改AndroidMainfest.xml中MainActivity的启动模式如下：

```
<activity android:name=".MainActivity" android:launchMode="singleTop">
```

此时再次运行程序，点击按钮进行跳转，可以看到屏幕没有切换动作，Android Monitor控制台的logcat中也没有新建实例对象的信息。退出程序只需要点击1次返回键。

如果修改代码，则MainActivity中的跳转为

```
Intent intent=new Intent(MainActivity.this, SecondActivity.class);
```

SecondActivity中的跳转为

```
Intent intent=new Intent(SecondActivity.this, MainActivity.class);
```

此时，由于第一次跳转后SecondActivity实例对象位于栈顶，MainActivity实例对象位于其后，第二次跳转完成后，将会再新建一个MainActivity实例对象，栈结构如图3.28所示。

| |
|---|
| MainActivity实例 |
| SecondActivity实例 |
| MainActivity实例 |

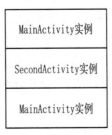

图3.28　singleTop启动模式示意图

### 3.5.3 singleTask 模式

singleTask 模式下启动 Activity，系统首先检查栈中是否存在该 Activity 的实例对象，如果存在，则直接返回该实例对象，并把在这个实例对象之上的其他对象全部清除出栈，保证该对象处于顶端。如果当前栈中没有该 Activity 的实例对象，则新建一个对象进入栈的顶端。

在上例中，如果修改 AndroidMainfest.xml 中 MainActivity 的启动模式如下：

<activity android:name=".MainActivity" android:launchMode="singleTask">

MainActivity 中的跳转为

Intent intent=new Intent(MainActivity.this, SecondActivity.class);

SecondActivity 中的跳转为

Intent intent=new Intent(SecondActivity.this, MainActivity.class);

则两次跳转后，栈中只剩下第一次启动时的 MainActivity 实例，因为栈中有了 MainActivity 实例对象，该对象直接被推到栈顶，SecondActivity 实例被清除出栈，如图 3.29 所示。

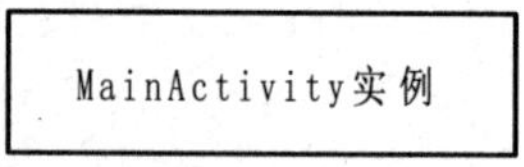

MainActivity 实例

图 3.29　singleTask 启动模式示意图

### 3.5.4 singleInstance 模式

singleInstance 模式下启动的 Activity 会产生一个新的栈来单独管理这个 Activity 实例对象，所有指向该 Activity 的跳转都共享这个实例对象，后续请求均不会创建它的新实例对象，即实现了 Activity 实例对象的共享。

在上例中，如果修改 AndroidMainfest.xml 中 MainActivity 的启动模式如下：

<activity android:name=".MainActivity" android:launchMode="singleInstance">

则两次跳转后，当前的 MainActivity 实例依然是第一次创建出的，第二次不会创建新的实例对象。栈结构可以理解如下，MainActivity 实例在一个单独的栈中，如图 3.30 所示。

MainActivity 实例　　SecondActivity 实例

图 3.30　singleInstance 启动模式示意图

在应用开发中，Activity 的这四种启动模式各自有着一些应用场景，例如推送应用可以设置成 singleTop 模式，减少重复创建对象；程序入口主界面可以设置成 singleTask 模式，这样每次返回主界面时，都可以回到第一次打开的主界面上，同时清空其他界面，减少内存消耗；一些独立的应用如计时提醒等可以设置为 singleInstance 模式，保证每次打开的都是同一个实例对象。

## 3.6 关于Context的说明

在之前的代码中多次见到了Context,例如:

(1) 语法声明:Toast.makeText(Context context, CharSequence text, int duration);

使用:Toast toast = Toast.makeText(MainActivity.this,"提示",Toast.LENGTH_SHORT);

这里Context传入的参数是MainActivity.this。

(2) 构造方法定义:

```
public ContactsAdapter(Context context, List<ContactsBean> list) {
    this.context = context;
    this.list = list;
}
```

在MainActivity中创建对象时,

new ContactsAdapter(this,contactsBeanList);

这里Context传入的参数是this。

(3) 将布局文件实例化成View对象时,

View view = LayoutInflater.from(viewGroup.getContext()).inflate(R.layout.item_contacts, viewGroup, false);

这里使用了getContext()方法获取当前组件的Context对象。

Context在Android应用中大量存在,它表示当前应用的上下文环境。Android应用及基础组件都是在一个特定的上下文环境中才能运行,Context是维持Android程序中各组件能够正常工作的一个核心功能类,Context是一个抽象类,它的类结构如图3.31所示。

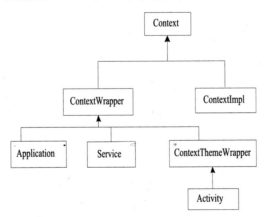

图3.31 Context类结构图

由类图可以看到,Application、Activity、Service等都是Context的子类,这些对象在初始化时都会被分配一个独立的上下文环境,通过这个上下文环境可以访问和使用这些对象自身的资源。例如:

(1) Toast在创建时,需要依赖一个Activity上下文环境,如果在Activity自身中,则可以写为

Toast toast = Toast.makeText(this,"提示",Toast.LENGTH_SHORT);

如果是在Activity的内部类或者内部接口实现中,则需要写为

Toast toast = Toast.makeText(MainActivity.this,"提示",Toast.LENGTH_SHORT);

这里的this和MainActivity.this都表示当前Activity的上下文环境。

(2) 在代码中创建View组件对象时,它需要依赖一个Context上下文环境,对View组件来说,一般是一个Activity上下文环境,例如在MainActivity的onCreate方法中创建一个按钮对象可以写为

Button button = new Button(this);

(3) 一个View对象可以通过自身的方法获得Context对象,其语法为view.getContext( ),例如:

viewGroup.getContext();

(4) 如果要获取整个应用Application的上下文环境,则可以在Activity中通过如下方法实现:

this.getApplicationContext();

# 第4章 项目主要界面设计与实现

一款典型的应用软件界面包括自动跳转的欢迎页、滑屏引导页、主功能页及各个功能界面。在项目开发中,可以通过计时器或者 Handler 的方式实现欢迎页自动跳转,通过 ViewPager 实现引导页滑屏,通过 Fragment 提供更加模块化的主功能页实现与管理,通过主题和样式的设置来保证软件风格统一。合理选择使用优秀的开源项目插件既可以提升项目开发效率,又可以实现更丰富的应用功能。

## 4.1 自动跳转的欢迎页

欢迎界面,也可以称为闪屏(Splash Screen),是一个应用程序启动时展现给用户的第一个界面,其作用原本是为了在程序加载的过程中给用户一个"友好的"等待界面,但是现在越来越多地用于展示 Logo 和宣传口号,推广品牌形象,也有一些程序会在欢迎界面发布广告,让用户先看广告。欢迎界面一般会在 3 到 10 秒自动跳转,一些广告形式的欢迎页可能时间会长些,这时建议提供一个"跳过"按钮,避免用户等待太长时间。

### 4.1.1 设置欢迎页为启动项——修改 Activity 启动顺序

**[项目案例3] 创建欢迎页并设置为启动项**

在项目 AccountBook 中新建一个"Empty Activity"(项目目录列表的"java"→包名 "cc.turbosnail.accountbook"上点击右键,选择"New"→"Activity"→"Empty Activity"),命名为 WelcomeActivity,然后打开项目目录 manifests 中的 AndroidManifest.xml 配置文件,把 <intent-filter>标签从".MainActivity"剪切到".WelcomeActivity"中,修改如下:

```
<activity android:name=".MainActivity"></activity>
<activity android:name=".WelcomeActivity">
    <intent-filter>
        <action android:name="android.intent.action.MAIN"/>
        <category android:name="android.intent.category.LAUNCHER"/>
```

```
        </intent-filter>
    </activity>
```

此时,把<intent-filter>过滤器添加给了 WelcomeActivity,并通过<action>标签声明它为启动 Activity,通过<category>标签声明它被添加到桌面启动图标列表中。程序启动时,".WelcomeActivity"成为启动 Activity,而不再是".MainActivity"。

### 4.1.2 欢迎页的实现——主题(Theme)、样式(Style)

欢迎页一般通过全屏的方式来展示内容,全屏效果可以基于主题来实现。主题是针对 Activity 或者 Application 设置外观效果的一个属性集合。它通过<style>标签定义,内容由<item>标签声明,通过 Activity 或者 Application 的"android:theme"属性引用。Android 提供了大量主题可以直接使用,如对 Application 默认设置的主题 android:theme="@style/AppTheme"。

**[项目案例4] 为欢迎页自定义主题,实现全屏显示**

(1) 打开 AccountBook 项目目录"res"→"values"中的 styles.xml 文件,添加自定义主题"WelcomeTheme"如下:

```
<style name="WelcomeTheme">
    <item name="android:background">#FFF</item>
    <item name="windowNoTitle">true</item>   <!--去掉标题栏,兼容v7包-->
    <item name="android:windowNoTitle">true</item>
    <item name="windowActionBar">false</item>   <!--不使用导航栏-->
    <item name="android:windowActionBar">false</item>
    <item name="android:windowFullscreen">true</item>   <!--设置全屏-->
    <item name="android:windowContentOverlay">@null</item>   <!--启动时没有背景遮盖-->
</style>
```

(2) 在项目目录 manifests 的 AndroidManifest.xml 配置文件中设置 WelcomeActivity 使用主题:

```
<activity android:name=".WelcomeActivity" android:theme="@style/WelcomeTheme">
```

(3) 在 activity_welcome.xml 中添加一个<ImageView>元素,设置其图片源为 Logo 图片"ic_logo.png":app:srcCompat="@drawable/ic_logo"。再添加一个<TextView>元素在<ImageView>下方,用来显示软件名称"随身记账本":android:text="随身记账本"。

运行程序,欢迎页 WelcomeActivity 成为启动时的第一个界面,此时当前界面背景色为白色,没有标题栏,全屏显示,效果如图4.1所示。

图4.1 欢迎页

除了通过主题设置欢迎页显示外观之外，还可以通过如下两种方式实现欢迎页的显示效果：

（1）在AndroidManifest.xml中为Activity添加系统自带的主题如下：

\<Activity android:name=".WelcomeActivity" android:theme="@style/Theme.AppCompat.NoActionBar"\>

（2）通过Java代码设置全屏。在Activity的setContentView语句之前添加如下语句：

```
//去除标题栏
supportRequestWindowFeature(Window.FEATURE_NO_TITLE);
//去除状态栏
getWindow().setFlags(WindowManager.LayoutParams.FLAG_FULLSCREEN,
        WindowManager.LayoutParams.FLAG_FULLSCREEN);
```

与主题相对应，样式是针对View等UI组件设置外观效果的一个属性集合。它也是通过\<style\>标签定义的，内容由\<item\>标签声明，不同的是它通过UI组件的"style"属性引用，如style="@style/myStyle"。我们可以把项目中一些组件的公共属性定义成样式文件供所有相关组件使用，这样既可以统一样式风格，又可以减少重复代码。

【例4.1】为按钮自定义样式。

（1）在Android Studio中新建一个项目，项目名为StyleDemo。打开项目目录"res"→"values"中的"style.xml"文件，添加一个\<style\>如下：

```
<style name="BtnStyle">
    <item name="android:layout_width">100dp</item>
    <item name="android:layout_height">wrap_content</item>
    <item name="android:textSize">20sp</item>
    <item name="android:textColor">#FFF</item>
    <!--背景使用了<selector>,请参见本书第2章,在项目中添加btn_selector.xml、btn_bg_normal.xml、btn_bg_pressed.xml文件,以及自定义颜色-->
```

        <item name="android:background">@drawable/btn_selector</item>
    </style>

（2）在布局文件 activity_main.xml 中添加 2 个按钮，在属性声明中均添加语句：style="@style/BtnStyle"，使用自定义样式 BtnStyle，代码如下：

```xml
<?xml version="1.0" encoding="utf-8"?>
<android.support.constraint.ConstraintLayout xmlns:android="http://schemas.android.com/apk/res/android"
    xmlns:app="http://schemas.android.com/apk/res-auto"
    xmlns:tools="http://schemas.android.com/tools"
    android:layout_width="match_parent"
    android:layout_height="match_parent"
    tools:context=".MainActivity">
    <Button
        android:id="@+id/btn_left_main"
        style="@style/BtnStyle"
        android:layout_marginTop="48dp"
        android:text="按钮一"
        app:layout_constraintLeft_toLeftOf="parent"
        app:layout_constraintRight_toLeftOf="@+id/btn_right_main"
        app:layout_constraintTop_toTopOf="parent"/>
    <Button
        android:id="@+id/btn_right_main"
        style="@style/BtnStyle"
        android:layout_marginTop="48dp"
        android:text="按钮二"
        app:layout_constraintLeft_toRightOf="@+id/btn_left_main"
        app:layout_constraintRight_toRightOf="parent"
        app:layout_constraintTop_toTopOf="parent"/>
</android.support.constraint.ConstraintLayout>
```

运行程序，如图 4.2 所示，可以看到两个按钮具有相同的样式：圆角、蓝色背景，字体和大小也相同。

图 4.2　为按钮设置样式

Android系统提供了大量的主题,也提供了丰富的<item>属性供开发者选用,在开发中可以根据需要查阅API文档选取合适的主题和样式属性。

### 4.1.3 欢迎界面的跳转——计时器、Handler

欢迎界面在3~10秒内应该自动跳转,这个功能可以通过定时器、Handler延时通知、Java线程休眠方法等多种方式实现。

(1)通过定时器跳转。定时器任务是Java线程的一种实现形式,一些简单的定时任务可以通过它来实现。定时器任务由Timer、TimerTask两个类配合完成,Timer是定时器,用来在后台线程按时间安排执行指定的任务。任务则由TimerTask子类来表示,其接口中的run方法被实现后,将在任务启动时被调用执行。

**[项目案例5.1] 基于定时器实现欢迎页定时跳转**

```
package cc.turbosnail.accountbook;
import android.content.Intent;
import android.support.v7.app.AppCompatActivity;
import android.os.Bundle;
import java.util.Timer;
import java.util.TimerTask;
public class WelcomeActivity extends AppCompatActivity {
    @Override
    protected void onCreate(Bundle savedInstanceState) {
        super.onCreate(savedInstanceState);
        setContentView(R.layout.activity_welcome);
        Timer timer = new Timer();//创建定时器
        /*实现TimerTask接口,任务在run方法中*/
        TimerTask timerTask = new TimerTask() {
            @Override
            public void run() {
                Intent intent = new Intent(WelcomeActivity.this, MainActivity.class);//根据需要跳转
                startActivity(intent);
                WelcomeActivity.this.finish();
            }
        };
        timer.schedule(timerTask, 1000 * 3);//定时器延迟3秒启动,执行任务
    }
}
```

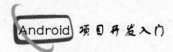

（2）通过 Handler 跳转。通过消息延时发送的方法 sendEmptyMessageDelayed 即可以实现通过 Handler 的跳转。当需要从网络端下载广告并更新界面显示，且加载大量数据时，可以将相关操作封装到子线程中，根据这些操作通知主线程修改界面内容，完成界面跳转。

**[项目案例5.2] 基于 Handler 实现欢迎页定时跳转**

```
package cc.turbosnail.accountbook;
import android.content.Intent;
import android.os.Handler;
import android.os.Message;
import android.support.v7.app.AppCompatActivity;
import android.os.Bundle;
public class WelcomeActivity extends AppCompatActivity {
    private Handler handler = new Handler() {
        @Override
        public void handleMessage(Message msg) {
            super.handleMessage(msg);
            Intent intent = new Intent(WelcomeActivity.this, MainActivity.class);//收到消息后跳转
            startActivity(intent);
            WelcomeActivity.this.finish();
        }
    };
    @Override
    protected void onCreate(Bundle savedInstanceState) {
        super.onCreate(savedInstanceState);
        setContentView(R.layout.activity_welcome);
        handler.sendEmptyMessageDelayed(0, 3000);//延迟3秒后发送一个空消息
    }
}
```

## 4.2 滑屏引导页——ViewPager

应用程序安装后第一次启动运行时，一般会从欢迎页跳转到引导页。引导页有多个界面，用来介绍软件功能、指导用户操作等，用户通过滑屏的形式滑动查看每个界面，在最后一个界面上会有一个"开始"按钮，用户点击后，程序进入下一个页面，例如登录界面或者功能主页面。当程序再次启动时，引导页一般不再显示。

引导页实现的关键是多个屏幕界面进行切换,Android 提供了 UI 组件 ViewPager 来完成滑动切换屏幕的功能。ViewPager 继承了 ViewGroup 类,即它可以作为容器添加要显示的 View 组件,但是 ViewPager 在添加 View 时需要使用 PagerAdapter 适配器,其实现过程为:

(1) 在布局中添加 ViewPager,标签为<android.support.v4.view.ViewPager>,可以设置它的各种属性,但是此时 ViewPager 中没有数据可以显示。

(2) 创建要显示的每一个 View,如为每一屏界面编写独立的 XML 布局文件;

(3) 自定义适配器类,该类是 PagerAdapter 适配器的子类,重写必需的方法,并定义接收数据源的方法(一般可以用一个自定义带参数的构造方法,参数用 List 型)。必须重写的四个方法是:

☞ public int getCount(),获取滑动组件的数量。

☞ public Object instantiateItem(ViewGroup container, int position),向 container 中根据滑动组件在列表中的位置 position 来添加滑动组件,返回值是滑动组件的 id,类型是 Object 型。

☞ public boolean isViewFromObject(View view, Object object),在滑动时判断 id 是否与滑动组件对应。其作用是决定一个页面 view 是否与 instantiateItem(ViewGroup, int)方法返回的具体 key 对象相关联。

☞ public void destroyItem(ViewGroup container, int position, Object object),把超出缓存范围的滑动组件移除。

(4) 构建 View 列表,以便于一次把所有 View 传递给适配器。例如:在 Activity 中通过 LayoutInflater 对象的 inflate 方法得到各 XML 布局文件的 View 型对象,把这些对象放置在一个 List 中,构成一个滑动组件列表。

注意:LayoutInflater 是用来访问 res/layout 目录下的 xml 布局文件的,findViewById()是用来访问 xml 布局文件里面具体的 View 组件的。

(5) 创建 PagerAdapter 适配器对象,把 View 列表传给它,即把滑动组件列表作为数据源传给适配器对象,由适配器对象进行统一管理。

(6) 获取 ViewPager 对象,然后通过 setAdapter 方法给它设置适配器。

[例 4.2] ViewPager 滑屏应用示例。创建一个带有 3 个滑屏的展示界面。

(1) 在 Android Studio 中新建一个项目,项目名为 ViewPagerDemo。

(2) 在 activity_main.xml 布局中添加<ViewPager>标签,代码如下:

```
<?xml version="1.0" encoding="utf-8"?>
<android.support.constraint.ConstraintLayout xmlns:android="http://schemas.android.com/apk/res/android"
    xmlns:app="http://schemas.android.com/apk/res-auto"
    xmlns:tools="http://schemas.android.com/tools"
    android:layout_width="match_parent"
    android:layout_height="match_parent"
```

```xml
        tools:context=".MainActivity">
        <android.support.v4.view.ViewPager
            android:id="@+id/viewpager"
            android:layout_width="match_parent"
            android:layout_height="match_parent">
        </android.support.v4.view.ViewPager>
</android.support.constraint.ConstraintLayout>
```

(3) 在项目目录"res"->"layout"上点击右键，分别选择"New"→"XML"→"Layout XML File"新建三个XML布局文件：guide_one.xml、guide_two.xml、guide_three.xml，新建的布局文件默认为线性布局，为区分这三个布局文件，分别给它们设置不同的背景颜色，并添加一个不同文字内容的文本标签。代码如下：

guide_one.xml代码清单：

```xml
<?xml version="1.0" encoding="utf-8"?>
<LinearLayout xmlns:android="http://schemas.android.com/apk/res/android"
    android:layout_width="match_parent"
    android:layout_height="match_parent"
    android:orientation="vertical"
    android:gravity="center"
    android:background="#F00">
    <TextView
        android:layout_width="match_parent"
        android:layout_height="wrap_content"
        android:text="页面一"
        android:textColor="#000"
        android:textSize="30sp"
        android:gravity="center"/>
</LinearLayout>
```

guide_two.xml、guide_three.xml代码内容与guide_one.xml相似，只需要修改属性android:background为"#0F0""#00F"，并修改android:text属性为"页面二""页面三"即可。

（4）创建适配器类。在项目目录"java"→包名"cc.turbosnail.viewpagerdemo"上点击右键，新建一个"Java Class"，类名为MyPagerAdapter，如图4.3所示。

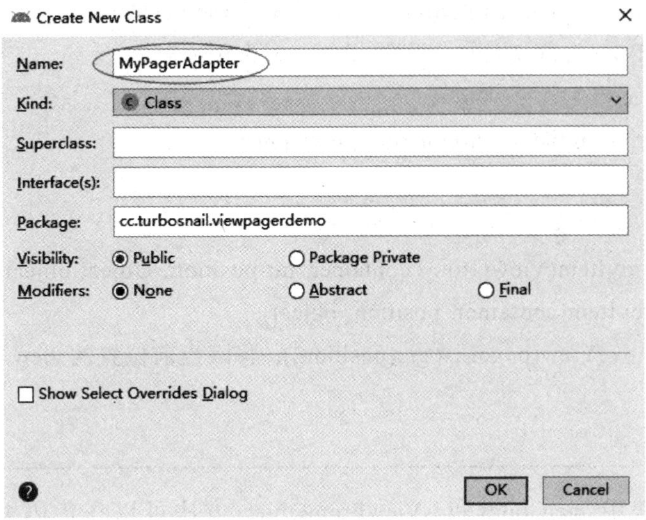

图 4.3　新建 Java 类

　　MyPagerAdapter 作为 PagerAdapter 的子类,自定义一个构造方法,用来接收界面列表对象,并根据接收的界面列表对象重写四个对应的方法,以实现滑屏效果。

```
package cc.turbosnail.viewpagerdemo;
import android.support.v4.view.PagerAdapter;
import android.view.View;
import android.view.ViewGroup;
import java.util.ArrayList;
public class MyPagerAdapter extends PagerAdapter {
    private ArrayList<View> pageList = null;//接收构造方法中传来的view列表
    public MyPagerAdapter(ArrayList<View>pageList) {
        this.pageList = pageList;
    }
    @Override
    public int getCount() {
        //return 0;
        return pageList.size();//返回view列表的大小,即view的数目
    }
    @Override
    public boolean isViewFromObject(View view, Object object) {
        //return false;
        return view == object;//返回当前显示的view是否与列表中的对象一致
    }
    @Override
    public Object instantiateItem(ViewGroup container, int position) {
```

```java
            View pageView = pageList.get(position);//获取当前位置的view
            container.addView(pageView);//设置当前view为显示对象
            return pageView;
            //return super.instantiateItem(container, position);
        }
        @Override
        public void destroyItem(ViewGroup container, int position, Object object) {
            // super.destroyItem(container, position, object);
            container.removeView(pageList.get(position));//移除当前位置的view
        }
    }
```

四个重写方法中，getCount 和 isViewFromObject 方法可以在代码编写时通过"ALT+Enter"键直接导入，instantiateItem 和 destroyItem 方法需要手动添加或者通过菜单栏中的"code"→"Override Methods"导入。本例以注释的形式保留了自动生成的代码部分，便于读者参照对比，理解适配器的工作过程。

（5）在 MainActivity 类中加载布局文件，添加适配器。代码如下：

```java
package cc.turbosnail.viewpagerdemo;
import android.support.v4.view.ViewPager;
import android.support.v7.app.AppCompatActivity;
import android.os.Bundle;
import android.view.LayoutInflater;
import android.view.View;
import java.util.ArrayList;
public class MainActivity extends AppCompatActivity {
    private ViewPager viewPager;
    private View viewOne, viewTwo, viewThree;//三个滑动界面对应的三个View对象
    private ArrayList<View> viewList = null;//存放三个滑动界面对象
    private MyPagerAdapter mAdapter = null;
    @Override
    protected void onCreate(Bundle savedInstanceState) {
        super.onCreate(savedInstanceState);
        setContentView(R.layout.activity_main);
        viewPager = findViewById(R.id.viewpager);
        /**获取三个布局文件,生成View对象**/
        LayoutInflater inflater = getLayoutInflater();
        viewOne = inflater.inflate(R.layout.guide_one, null);
        viewTwo = inflater.inflate(R.layout.guide_two, null);
```

viewThree = inflater.inflate(R.layout.guide_three, null);
viewList = new ArrayList();
viewList.add(viewOne);//构建显示组件的列表
viewList.add(viewTwo);
viewList.add(viewThree);
mAdapter = new MyPagerAdapter(viewList);
viewPager.setAdapter(mAdapter);//添加适配器
    }
}

运行程序,左右滑动屏幕,可以看到界面效果如图4.4所示。

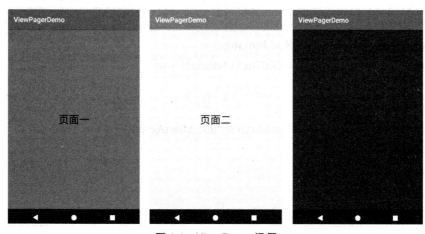

图4.4  ViewPager滑屏

[项目案例6] 为AccountBook项目添加滑屏引导页

(1) 打开AccountBook项目,新建一个Activity,命名为GuideActivity。在项目目录列表的"java"→包名"cc.turbosnail.accountbook"上点击右键,选择"New"→"Activity"→"Empty Activity"。同时,可以在项目目录manifests的AndroidManifest.xml配置文件中为GuideActivity设置主题,让它全屏显示。

(2) 在activity_guide.xml布局中添加<ViewPager>标签。

(3) 在项目目录"res"→"layout"上点击右键,选择"New"→"XML"→"Layout XML File",分别新建三个XML布局文件作为引导界面,并根据需要为这三个文件添加显示内容。布局文件可以分别命名为guide_one.xml、guide_two.xml、guide_three.xml,其中,guide_three.xml除了自身显示的内容外,需要再添加一个跳转按钮如下:

```
<Button
    android:id="@+id/btn_in"
    android:layout_width="match_parent"
    android:layout_height="wrap_content"
```

```
android:text="开始体验"
android:textColor="#000"
android:textSize="30sp"
android:gravity="center"/>
```

（4）创建适配器类。类命名为 GuidePagerAdapter，并实现适配器代码。

（5）在 GuideActivity 类中加载布局文件，添加适配器。同时，在 onCreate 方法中通过 viewThree.findViewById()方法获得按钮对象，并添加事件监听进行跳转。

注意，此时的按钮由于来自于 guide_three.xml，guide_three.xml 已经通过 inflate 方法加载给了 View 类型的对象 viewThree，所以要使用 viewThree.findViewById()，不能直接使用 findViewById()。代码示例如下：

```
viewThree = inflater.inflate(R.layout.guide_three, null);
btnIn = viewThree.findViewById(R.id.btn_in);
btnIn.setOnClickListener(new View.OnClickListener() {
    @Override
    public void onClick(View v) {
        Intent intent = new Intent(GuideActivity.this,MianActivity.class);
        startActivity(intent);
        GuideActivity.this.finish();
    }
});
```

运行程序，可以看到引导页首先启动起来，当滑屏到第三个界面上时，可以通过"开始体验"按钮跳转到 MianActivity。

## 4.3 主功能页——Fragment

主功能页是呈现给用户的第一个工作界面，例如打开微信 APP 后看到的消息列表界面。一般的应用软件都会有多个工作界面，为方便操作，通过顶部或者底部导航供用户选择，例如打开微信后，我们会看到"微信""通信录""发现""我"四个底部导航按钮。如果在工作界面切换时，底部导航不用切换，即只让导航栏以外的内容部分进行切换，且各个内容部分又可以像 Activity 一样具有独立的生命周期，以实现各种复杂的业务逻辑功能，这样既可以减少使用多个代码结构重复的 Activity，又可以基于模块化的方式降低界面的复杂度。Android 提供了 Fragment 来让开发者实现这样的效果。

### 4.3.1 Fragment

Fragment 是一个用来对界面进行分块管理的组件，它是"小型的 Activity"，像 Activity 一样具有独立的生命周期。通过 Fragment 可以有效实现界面在布局结构、功能响应等方面的模块化开发,降低界面设计的复杂度,提升代码复用性,界面更新和事件响应更迅速。

Fragment从创建到销毁的过程与Activity类似,Android也定义了生命周期各个阶段对应的方法供开发者调用,Fragment完整的生命周期如图4.5所示。

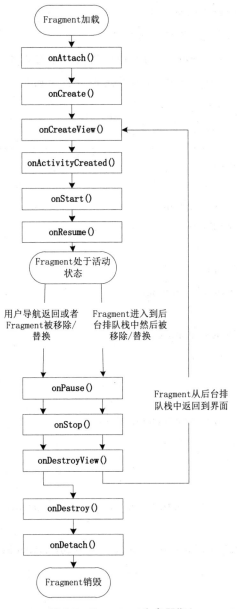

图4.5　Fragment生命周期

在这个生命周期的过程中,Fragment中对应的方法将被调用执行,和Activity类似,这些方法在不同的生命周期阶段发挥作用,主要包括如下方法:

☞ onAttach(Context),绑定Fragment到Activit时调用。

☞ onCreate(Bundle),初始化Fragment时调用。

☞ onCreateView(LayoutInflater,ViewGroup,Bundle),Fragment创建界面视图时调用。

- onActivityCreated(Bundle),当 Activity 中的 onCreate 方法执行完成后调用。
- onStart()Fragment,Fragment 启动时调用,此时 Fragment 可见。
- onResume(),Fragment 获取焦点时调用。
- onPause(),Fragment 失去焦点时调用。
- onStop(),Fragment 停止不可见时调用。
- onDestroyView(),Fragment 中的界面视图被移除时调用。
- onDetach(),Fragment 与 Activity 解除关联分离时回调。
- onDestroy(),Fragment 销毁时调用。

Fragment 生命周期对应的代码结构如下:

```java
import android.content.Context;
import android.os.Bundle;
import android.support.v4.app.Fragment;
import android.view.LayoutInflater;
import android.view.View;
import android.view.ViewGroup;
public class FragmentTest extends Fragment {
    @Override
    public void onAttach(Context context) {
        super.onAttach(context);
    }
    @Override
    public void onCreate(Bundle savedInstanceState) {
        super.onCreate(savedInstanceState);
    }
    @Override
    public View onCreateView(LayoutInflater inflater, ViewGroup container, Bundle savedInstanceState) {
        return super.onCreateView(inflater, container, savedInstanceState);
    }
    @Override
    public void onActivityCreated(Bundle savedInstanceState) {
        super.onActivityCreated(savedInstanceState);
    }
    @Override
    public void onViewStateRestored(Bundle savedInstanceState) {
        super.onViewStateRestored(savedInstanceState);
    }
```

```java
    @Override
    public void onStart() {
        super.onStart();
    }
    @Override
    public void onResume() {
        super.onResume();
    }
    @Override
    public void onPause() {
        super.onPause();
    }
    @Override
    public void onStop() {
        super.onStop();
    }
    @Override
    public void onDetach() {
        super.onDetach();
    }
    @Override
    public void onDestroyView() {
        super.onDestroyView();
    }
    @Override
    public void onDestroy() {
        super.onDestroy();
    }
}
```

Fragment虽然有自己独立的生命周期，但是不能单独使用，必须嵌套在Activity中使用，其生命周期受到宿主Activity生命周期的影响，如果宿主Activity被销毁，它也会被销毁。Fragment和宿主Activity之间可以互相通信，主要的方法如下：

- getActivity().findViewById(R.id.组件id)：在Fragment代码中获得宿主Activity中的组件。
- getSupportFragmentManager().findFragmentByid(R.id.Fragment的id)：在Activity代码中获得Fragment对象。

Fragment的创建步骤如下：

（1）创建Fragment的布局文件，如基于XML的布局文件。

（2）自定义Fragment类，继承Fragment，重写onCreateView()方法，在该方法中加载Fragment的布局文件，并返回view对象。

（3）在Activity的布局文件中添加<fragment>标签，name属性是自定义Fragment类的"包名+类名"，如android:name="cc.turbosnail.fragmentdemo.FragmentTitle"。

[例4.3] 创建Fragment示例。在一个Activity中分别创建两个Fragment。

（1）在Android Studio中新建一个项目，项目名为FragmentDemo。

（2）为Fragment创建布局文件。在项目目录"res"→"layout"上点击右键，选择"New"→"XML"→"Layout XML File"，分别新建两个XML布局文件fragment_title.xml和fragment_content.xml，并实现布局内容如下：

fragment_title.xml用来模拟一个章节目录，代码清单如下：

```xml
<?xml version="1.0" encoding="utf-8"?>
<LinearLayout xmlns:android="http://schemas.android.com/apk/res/android"
    android:layout_width="match_parent"
    android:layout_height="match_parent"
    android:orientation="vertical"
    android:gravity="left"
    android:background="#317ef3">
    <TextView
        android:id="@+id/tv_ch1"
        android:layout_width="wrap_content"
        android:layout_height="wrap_content"
        android:text="第一章"
        android:textColor="#FFF"
        android:textSize="20sp"
        android:lineSpacingMultiplier="1.5"/>
    <TextView
        android:id="@+id/tv_ch2"
        android:layout_width="wrap_content"
        android:layout_height="wrap_content"
        android:text="第二章"
        android:textColor="#FFF"
        android:textSize="20sp"
        android:lineSpacingMultiplier="1.5"/>
</LinearLayout>
```

fragment_content.xml用来显示章节的内容，代码清单如下：

```xml
<?xml version="1.0" encoding="utf-8"?>
<LinearLayout xmlns:android="http://schemas.android.com/apk/res/android"
```

```xml
    android:layout_width="match_parent"
    android:layout_height="match_parent"
    android:orientation="vertical"
    android:background="#317ef3">
    <TextView
        android:id="@+id/tv_content"
        android:layout_width="wrap_content"
        android:layout_height="wrap_content"
        android:gravity="left"
        android:text="请点击目录阅读"
        android:textSize="20sp"
        android:textStyle="bold"
        android:lineSpacingMultiplier="1.5"/>
</LinearLayout>
```

(3) 创建 Fragment 类。在项目目录"java"->包名"cc.turbosnail.fragmentdemo"上点击右键，选择"New"→"Java Class"，分别新建两个 Java 类：FragmentTitle 和 FragmentContent，在代码中实现 Fragment 的子类，并在 onCreateView 方法中加载对应的 XML 布局文件。两个类的代码清单分别如下：

FragmentTitle.java:
```java
package cc.turbosnail.fragmentdemo;
import android.os.Bundle;
import android.support.v4.app.Fragment;
import android.view.LayoutInflater;
import android.view.View;
import android.view.ViewGroup;
public class FragmentTitle extends Fragment {
    @Override
    public View onCreateView(LayoutInflater inflater, ViewGroup container, Bundle savedInstanceState) {
        View view = inflater.inflate(R.layout.fragment_title, null);
        return view;
    }
}
```

FragmentContent.java 中为界面上的 Fragment 对象添加了一个点击事件监听器，当用户点击这个 Fragment 对象时，则会弹出一个提示框，代码清单如下：

```java
package cc.turbosnail.fragmentdemo;
import android.os.Bundle;
import android.support.v4.app.Fragment;
import android.view.LayoutInflater;
import android.view.View;
import android.view.ViewGroup;
import android.widget.Toast;
public class FragmentContent extends Fragment {
    @Override
    public View onCreateView(LayoutInflater inflater, ViewGroup container, Bundle savedInstanceState) {
        View view = inflater.inflate(R.layout.fragment_content, null);
        view.setOnClickListener(new View.OnClickListener() {
            @Override
            public void onClick(View view) {
                Toast.makeText(getActivity(), "你点击了内容部分", Toast.LENGTH_SHORT).show();
            }
        });
        return view;
    }
}
```

（4）在Activity的布局文件activity_main.xml中添加<fragment>标签。添加两个上下排列的<fragment>标签，分别用来显示目录和内容，设置它们的name属性值是对应的类名，并设置id属性如下：

```xml
<fragment
    android:id="@+id/fg_title"
    android:name="cc.turbosnail.fragmentdemo.FragmentTitle"
    android:layout_width="match_parent"
    android:layout_height="60dp"
    ……/><!-- 此处省略了布局代码-->
<fragment
    android:id="@+id/fg_content"
    android:name="cc.turbosnail.fragmentdemo.FragmentContent"
    android:layout_width="match_parent"
    android:layout_height="440dp"
    ……/><!-- 此处省略了布局代码-->
```

（5）在 MainActivity 中对 Fragment 中的组件直接进行监听，并响应事件。监听 Fragment 中的标题文本标签，当第一个标签被点击时输出一段文字，当第二个标签被点击时输出另外一段文字，代码清单如下：

```java
package cc.turbosnail.fragmentdemo;
import android.support.v7.app.AppCompatActivity;
import android.os.Bundle;
import android.view.View;
import android.widget.TextView;
public class MainActivity extends AppCompatActivity implements View.OnClickListener {
    TextView tvCh1, tvCh2, tvContent;
    @Override
    protected void onCreate(Bundle savedInstanceState) {
        super.onCreate(savedInstanceState);
        setContentView(R.layout.activity_main);
        tvCh1 = findViewById(R.id.tv_ch1);//可以直接获取Fragment中的组件
        tvCh2 = findViewById(R.id.tv_ch2);
        tvContent = findViewById(R.id.tv_content);
        tvCh1.setOnClickListener(this);//为标题文本标签添加监视器
        tvCh2.setOnClickListener(this);
    }
    @Override
    public void onClick(View v) {
        switch (v.getId()) {
            case R.id.tv_ch1:
                tvContent.setText("当我还只有六岁的时候,在一本描写原始森林的名叫《真实的故事》的书中,看到了一幅精彩的插画,画的是一条蟒蛇正在吞食一只大野兽。");
                break;
            case R.id.tv_ch2:
                tvContent.setText("我就这样孤独地生活着,没有一个能真正谈得来的人,一直到六年前在撒哈拉沙漠上发生了那件事。");
                break;
            default:
                break;
        }
    }
}
```

运行程序，点击"标题1"和"标题2"，可以看到显示不同的内容，同时，点击下方的内容显示区域，还会弹出一个提示信息。运行效果如图4.6所示。

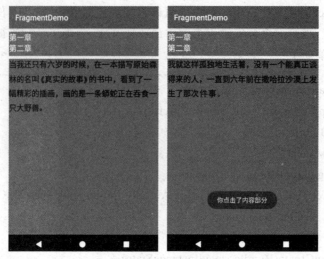

图4.6 Fragment简单示例

### 4.3.2 主页面实现——Fragment动态加载

**1. 动态加载Fragment**

当每一个工作界面都由Fragment构成时,就需要动态添加、显示、删除、隐藏对应的Fragment,以保证根据用户选择展现不同工作界面,这也是Fragment在应用开发中的主要用法。Activity通过FragmentManager来动态管理Fragment,实现Fragment的添加、隐藏等操作,具体步骤如下:

(1) 在Activity的布局文件如activity_main.xml中为Fragment添加一个容器布局,定义好id属性。建议使用FrameLayout布局。注意,此时不需要将Fragment添加到容器布局中。

```
<FrameLayout
    android:id="@+id/fl_main"
    android:layout_width="match_parent"
    android:layout_height="match_parent"/>
```

(2) 在Activity的Java代码中创建各个Fragment对象。这是标准的Java对象声明和创建操作。例如:

```
MainAccountFragment mainAccountFragment = new MainAccountFragment();
MainRecordFragment mainRecordFragment = new MainRecordFragment();
```

(3) 在Activity的Java代码中获取FragmentManager对象,然后利用进一步得到的FragmentTransaction对象来动态管理各个Fragment对象。相关操作必须放置在事务开始(beginTransaction)和提交(commit)之间。

添加多个Fragment对象示例代码如下:

```
FragmentManager manager = getSupportFragmentManager();
FragmentTransaction transaction = manager.beginTransaction();
ransaction.add(R.id.fl_main, mainAccountFragment);
transaction.add(R.id.fl_main, mainRecordFragment);
transaction.commit();
```

（4）用FragmentTransaction对象来完成在界面上动态显示/隐藏Fragment对象的操作，示例代码如下：

```
transaction = manager.beginTransaction();
transaction.show(mainAccountFragment);
transaction.hide(mainRecordFragment);
transaction.commit();
```

2. 项目主界面框架设计与实现

设计思想：

设计底部导航栏。把一个单选按钮组（RadioGroup、RadioButton）放置在主界面的底部，由三个单选按钮作为导航按钮，分别为"账目列表""记一笔""图表统计"。这三个栏目分别对应账目显示列表、记账、图标统计功能模块。当点击导航按钮时，切换到对应的Fragment上。

设计各个功能界面框架。每个功能界面的框架都实现成一个Fragment，在对应的XML布局文件中设计布局，在对应的Java代码中实现功能。界面切换时，通过在代码中向主界面动态添加或者隐藏Fragment对象，实现界面切换效果。

实现步骤：

**[项目案例7] 为AccountBook项目设计并实现带底部导航的主功能页面**

（1）设计主界面整体布局。

在项目目录"res"→"drawable"中创建两个shape文件用来表示选中和未选中状态的颜色：rb_bg_normal.xml、rb_bg_pressed.xml，创建一个selector文件用来引用颜色状态：rb_selector.xml（创建过程请参照第2章：按钮（Button））。

rb_bg_normal.xml代码清单：

```xml
<?xml version="1.0" encoding="utf-8"?>
<shape xmlns:android="http://schemas.android.com/apk/res/android">
    <solid android:color="#317ef3"/>
</shape>
```

rb_bg_pressed.xml代码清单：

```xml
<?xml version="1.0" encoding="utf-8"?>
<shape xmlns:android="http://schemas.android.com/apk/res/android">
```

```xml
        <solid android:color="#8eb9f5"/>
</shape>
```

rb_selector.xml代码清单:

```xml
<?xml version="1.0" encoding="utf-8"?>
<selector xmlns:android="http://schemas.android.com/apk/res/android">
    <item android:state_checked="true" android:drawable="@drawable/rb_bg_pressed"/>
    <item android:state_checked="false" android:drawable="@drawable/rb_bg_normal"/>
</selector>
```

在activity_main.xml主界面布局文件中,上部分添加一个<FrameLayout>标签,设置id和位置等属性,其高度占据屏幕主体,只为底部导航栏留下空间即可。底部导航栏为一个单选按钮组,放置三个单选按钮,通过android:button="@null"属性设置去掉选择框,通过android:background="@drawable/rb_selector"设置选中/未选中的颜色状态,通过android:layout_weight="1"设置它们平分位置,标题分别设置为"账目列表""记一笔""图表统计"。底部导航栏建议高度为56 dp。

```xml
<?xml version="1.0" encoding="utf-8"?>
<android.support.constraint.ConstraintLayout xmlns:android="http://schemas.android.com/apk/res/android"
    xmlns:app="http://schemas.android.com/apk/res-auto"
    xmlns:tools="http://schemas.android.com/tools"
    android:layout_width="match_parent"
    android:layout_height="match_parent"
    tools:context=".MainActivity">
    <FrameLayout
        android:id="@+id/fy_main"
        android:layout_width="match_parent"
        android:layout_height="0dp"
        app:layout_constraintBottom_toTopOf="@+id/rg_select"
        app:layout_constraintLeft_toLeftOf="parent"
        app:layout_constraintRight_toRightOf="parent"
        app:layout_constraintTop_toTopOf="parent"/>
    <!--导航栏高度建议为56dp-->
    <!--android:layout_marginRight="1dp"用来简单分隔按钮边界-->
    <RadioGroup
        android:id="@+id/rg_select"
        android:layout_width="match_parent"
        android:layout_height="56dp"
        android:orientation="horizontal"
```

```xml
        app:layout_constraintBottom_toBottomOf="parent"
        app:layout_constraintLeft_toLeftOf="parent"
        app:layout_constraintRight_toRightOf="parent">
        <RadioButton
            android:id="@+id/rb_account"
            android:layout_width="0dp"
            android:layout_height="match_parent"
            android:layout_marginRight="1dp"
            android:layout_weight="1"
            android:background="@drawable/rb_selector"
            android:button="@null"
            android:gravity="center"
            android:text="账目列表"
            android:textColor="#FFF"
            android:textSize="20sp"/>
        <RadioButton
            android:id="@+id/rb_write"
            android:layout_width="0dp"
            android:layout_height="match_parent"
            android:layout_marginRight="1dp"
            android:layout_weight="1"
            android:background="@drawable/rb_selector"
            android:button="@null"
            android:gravity="center"
            android:text="记一笔"
            android:textColor="#FFF"
            android:textSize="20sp"/>
        <RadioButton
            android:id="@+id/rb_chart"
            android:layout_width="0dp"
            android:layout_height="match_parent"
            android:layout_weight="1"
            android:background="@drawable/rb_selector"
            android:button="@null"
            android:gravity="center"
            android:text="图表分析"
            android:textColor="#FFF"
            android:textSize="20sp"/>
    </RadioGroup>
</android.support.constraint.ConstraintLayout>
```

（2）为Fragment创建布局文件。在项目目录"res"→"layout"上点击右键，选择"New"→"XML"→"Layout XML File"，分别新建三个XML布局文件：main_account.xml、main_record.xml、main_chart.xml。这三个文件将作为"账目列表""记一笔""图表统计"三个功能对应的布局文件。当前，可以用简单界面模拟工作界面，等到整体框架搭建完成后再进一步细化这三个工作界面。例如，只通过一个文本标签和背景颜色设置来标识第一个工作界面的main_account.xml代码清单如下：

```xml
<?xml version="1.0" encoding="utf-8"?>
<android.support.constraint.ConstraintLayout xmlns:android="http://schemas.android.com/apk/res/android"
    android:layout_width="match_parent"
    android:layout_height="match_parent">
    <TextView
        android:layout_width="match_parent"
        android:layout_height="match_parent"
        android:background="#F00"
        android:text="账目列表界面"
        android:textSize="20dp"/>
</android.support.constraint.ConstraintLayout>
```

（3）创建Fragment类，加载界面布局。分别创建三个Fragment类：MainAccountFragment、MainRecordFragment、MainChartFragment，分别加载各自的布局文件。例如，MainAccountFragment的代码清单如下：

```java
package cc.turbosnail.accountbook;
import android.os.Bundle;
import android.support.v4.app.Fragment;
import android.view.LayoutInflater;
import android.view.View;
import android.view.ViewGroup;
public class MainAccountFragment extends Fragment {
    public View onCreateView(LayoutInflater inflater, ViewGroup container,
                Bundle savedInstanceState) {
        View view = inflater.inflate(R.layout.main_account, container, false);
        return view;
    }
}
```

（4）在 MainActivity 中创建各个 Fragment 子类对象，并进行动态管理。通过 FragmentManager 添加所有 Fragment 子类对象到容器组件中，并根据用户在导航栏中的选择在事务中显示/隐藏 Fragment 子类对象，实现 Fragment 切换的效果。

```java
package cc.turbosnail.accountbook;
import android.support.v4.app.FragmentManager;
import android.support.v4.app.FragmentTransaction;
import android.support.v7.app.AppCompatActivity;
import android.os.Bundle;
import android.widget.RadioGroup;
public class MainActivity extends AppCompatActivity {
    private RadioGroup rgSelect;
    private FragmentManager manager;
    private FragmentTransaction transaction;
    private MainAccountFragment mainAccountFragment;//声明自定义的Fragment对象
    private MainRecordFragment mainRecordFragment;
    private MainChartFragment mainChartFragment;
    @Override
    protected void onCreate(Bundle savedInstanceState) {
        super.onCreate(savedInstanceState);
        setContentView(R.layout.activity_main);
        mainAccountFragment = new MainAccountFragment();
        mainRecordFragment = new MainRecordFragment();
        mainChartFragment = new MainChartFragment();
        manager = getSupportFragmentManager();
        transaction = manager.beginTransaction();//启动事务
        /*依次添加Fragment对象到容器组件fy_main中*/
        transaction.add(R.id.fy_main, mainAccountFragment);
        transaction.add(R.id.fy_main, mainRecordFragment);
        transaction.add(R.id.fy_main, mainChartFragment);
        /*显示一个Fragment对象,隐藏其他2个*/
        transaction.show(mainAccountFragment);
        transaction.hide(mainRecordFragment);
        transaction.hide(mainChartFragment);
        transaction.commit();//提交事务
        /*监听单选按钮组,在事务中根据不同按钮显示对应的Fragment对象,并隐藏其他两个,实现Fragment切换效果*/
        rgSelect = findViewById(R.id.rg_select);
        rgSelect.setOnCheckedChangeListener(new RadioGroup.OnCheckedChangeListener() {
```

```java
        @Override
        public void onCheckedChanged(RadioGroup group, int checkedId) {
            transaction = manager.beginTransaction();//启动事务
            switch (checkedId) {
                case R.id.rb_account:
                    transaction.show(mainAccountFragment);
                    transaction.hide(mainRecordFragment);
                    transaction.hide(mainChartFragment);
                    break;
                case R.id.rb_write:
                    transaction.hide(mainAccountFragment);
                    transaction.show(mainRecordFragment);
                    transaction.hide(mainChartFragment);
                    break;
                case R.id.rb_chart:
                    transaction.hide(mainAccountFragment);
                    transaction.hide(mainRecordFragment);
                    transaction.show(mainChartFragment);
                    break;
            }
            transaction.commit();//提交事务
        }
    });
}
}
```

运行程序,工作界面可以通过底部导航进行切换,效果如图4.7所示。

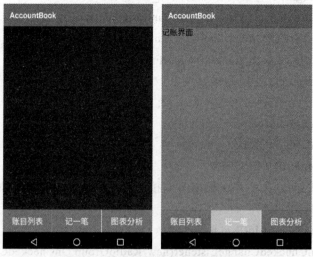

图4.7 随身记账本项目主界面框架

## 4.4 账目列表滑屏切换(ViewPager+Fragment)

在账目列表显示时,通过一个顶部导航,允许用户点击导航标签分类查看,如"全部""用餐""服装"等,同时,为提供更好的用户体验,多数应用软件提供滑屏导航的功能。此时,可以结合ViewPager+Fragment来实现这样的功能,既能通过导航按钮切换,又能通过滑屏切换界面,且每个界面模块都有独立的生命周期,方便实现各种业务逻辑。

### 4.4.1 ViewPager+Fragment

用Fragment作为ViewPager的显示组件,可以充分利用Fragment独立生命周期管理的优势实现模块化开发,而且能提供更好的用户体验效果,在应用开发中具有广泛的应用。其具体实现流程如下:

(1) 主界面设计。放置一个ViewPager作为Fragment的容器。

(2) 创建各个Fragment的布局文件、Java类。

(3) 创建FragmentPagerAdapter适配器类。用FragmentPagerAdapter适配器代替PagerAdapter适配器,这种方式使得代码更简洁,推荐使用。一个典型的FragmentPagerAdapter适配器可以定义如下:

```java
package cc.turbosnail.accountbook;
import android.support.v4.app.Fragment;
import android.support.v4.app.FragmentManager;
import android.support.v4.app.FragmentPagerAdapter;
import java.util.List;
public class AccountFragAdapter extends FragmentPagerAdapter {
  private List<Fragment> fragmentList;//自定义List用来接收多个Fragment对象列表
  /*在构造方法中添加一个 List<Fragment>参数接收数据*/
  public AccountFragAdapter(FragmentManager fm, List<Fragment> fragmentList) {
    super(fm);
    this.fragmentList = fragmentList;
  }
  @Override
  public Fragment getItem(int i) {
    //return null;
    return fragmentList.get(i);//重写,返回自定义List中的元素
  }
  @Override
  public int getCount() {
```

```
        //return 0;
           return fragmentList.size();//重写,返回自定义List的大小
    }
}
```

(4) 在主界面对应的 Java 代码中获取到 ViewPager 容器对象,创建用来滑屏显示的各个 Fragment 对象并添加到 List 列表中,然后以这个 List 列表作为参数之一创建适配器对象,为 ViewPager 容器对象添加适配器。示例代码如下:

```
viewPager = view.findViewById(R.id.vp_list_account);
rgSelect = view.findViewById(R.id.rg_select_account);
List<Fragment> fragmentList = new ArrayList();
fragmentList.add(new AccountAllFragment());
fragmentList.add(new AccountEatFragment());
fragmentList.add(new AccountClothesFragment());
fragmentList.add(new AccountTrafficFragment());
fragmentList.add(new AccountPlayFragment());
AccountFragAdapter adapter = new AccountFragAdapter(getChildFragmentManager(), fragmentList);
viewPager.setAdapter(adapter);//添加适配器
```

关于适配器构造方法中的第一个参数说明:当前的 Fragment 在另一个 Fragment 中,用 getChildFragmentManager( )方法获取 FragmentManager 对象;当前的 Fragment 如果是在一个 Activity 中,则使用 getFragmentManager( )方法获取 FragmentManager 对象。

### 4.4.2 带滑屏显示的账目列表

设计思想:

在项目案例中,当前的工作主界面是一个 Fragment,即用户点击屏幕下方导航按钮"账目"时显示的 MainAccountFragment 对象,它对应的布局文件是 main_account.xml。所以,可以把 viewPager 放置到这个 Fragment 的布局文件中,然后再在 viewPager 中添加多个 Fragment 对象,每个 Fragment 对象代表一种显示类型,例如用餐、交通、服装、娱乐等。同时,为了实现导航点击切换功能,可以在布局文件中通过单选按钮组(RadioGroup、RadioButton)设计一个顶部导航栏。具体的账目列表则通过 RecyclerView 来显示。整体布局如图 4.8 所示。

# 第 4 章 项目主要界面设计与实现

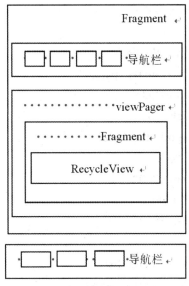

图 4.8 账目列表界面设计示意图

实现步骤：

**[项目案例 8]** 为 **AccountBook** 项目设计并实现带滑屏显示功能的账目显示功能

（1）基于 RecyclerView 设计和实现账目显示列表项。

在本项目中，每一个账目列表项要显示的内容包括照片、账目内容、账目描述、金额、时间等信息。设计布局文件 item_account.xml，代码清单如下：

```xml
<?xml version="1.0" encoding="utf-8"?>
<LinearLayout xmlns:android="http://schemas.android.com/apk/res/android"
    android:layout_width="match_parent"
    android:layout_height="60dp"
    android:orientation="horizontal"
    android:background="#FFF">
    <ImageView
        android:id="@+id/iv_type_list"
        android:layout_width="45dp"
        android:layout_height="45dp"
        android:layout_gravity="center_vertical"
        android:layout_marginLeft="10dp"
        android:src="@drawable/ic_icon"/>
    <LinearLayout
        android:layout_height="match_parent"
        android:layout_width="0dp"
        android:layout_weight="1"
```

```xml
            android:orientation="vertical">
            <TextView
                android:id="@+id/tv_title_list"
                android:layout_width="match_parent"
                android:layout_height="0dp"
                android:layout_weight="1"
                android:layout_marginLeft="10dp"
                android:text="暂无数据"
                android:gravity="center|left"
                android:textColor="#000"
                android:textSize="16sp"/>
            <TextView
                android:id="@+id/tv_explain_list"
                android:layout_width="match_parent"
                android:layout_height="0dp"
                android:layout_weight="1"
                android:layout_marginLeft="10dp"
                android:text="暂无数据"
                android:gravity="center|left"
                android:textColor="#999"
                android:textSize="16sp"/>
            <TextView
                android:id="@+id/tv_time_list"
                android:layout_width="match_parent"
                android:layout_height="0dp"
                android:layout_weight="1"
                android:layout_marginLeft="10dp"
                android:text="2018-11-26"
                android:gravity="center|left"
                android:textColor="#999"
                android:textSize="14sp"/>
        </LinearLayout>
        <TextView
            android:id="@+id/tv_money_list"
            android:layout_width="60dp"
            android:layout_height="60dp"
            android:layout_gravity="center_vertical"
            android:text="0元"
            android:gravity="center"
```

```
            android:textColor="#000"
            android:textSize="20sp"/>
</LinearLayout>
```

根据要显示的数据创建 JavaBean 类。创建一个 Java Class 用来存储账目信息，命名为 CostBean。可以在 Android Studio 的菜单"Code"→"Generate…"中直接生成对应的 set/get 方法，CostBean.java 部分代码如下：

```
package cc.turbosnail.accountbook;
public class CostBean {
    private int costID;//账目 id
    private int userID;//用户 id
    private float money;//金额
    private String type;//账目类型
    private String title;//账目内容
    private String explain;//账目详细描述
    private String time;//时间
    private String picPath;//图片路径
    public void setCostID(int costID) {
        this.costID = costID;
    }
    public int getCostID() {
        return costID;
    }
    public int getUserID() {
        return userID;
    }
    public void setUserID(int userID) {
        this.userID = userID;
    }
    ……此处省略其他成员变量的 set/get 方法,请读者自行补充。
}
```

创建 RecycleView 适配器。新建一个 Java Class,命名为 AccountItemAdapter,在项目目录"Gradle Scripts"→"build.gradle（Module:app）"中添加"implementation 'com.android.support:recyclerview-v7:28.0.0'",注意版本对应。然后在 Java 代码中导入包,并实现适配器如下：

```
package cc.turbosnail.accountbook;
import android.content.Context;
```

```java
import android.support.v7.widget.RecyclerView;//导入包
import android.view.LayoutInflater;
import android.view.View;
import android.view.ViewGroup;
import android.widget.ImageView;
import android.widget.TextView;
import java.util.List;
public class AccountItemAdapter extends RecyclerView.Adapter<AccountItemAdapter.ViewHolder> {
    private List<CostBean> list;
    private Context context;
    public AccountItemAdapter(Context context, List<CostBean> list) {
        this.context = context;
        this.list = list;
    }
    @Override
    public ViewHolder onCreateViewHolder(ViewGroup viewGroup, int i) {
        View view = LayoutInflater.from(viewGroup.getContext()).inflate(R.layout.account_item, viewGroup, false);
        ViewHolder viewHolder = new ViewHolder(view);
        return viewHolder;
    }
    @Override
    public void onBindViewHolder(ViewHolder viewHolder, int i) {
        CostBean bean = list.get(i);
        /*通过类型判断,为ivType设置不同的图片*/
        if (bean.getType().equals("用餐")) {
            viewHolder.ivType.setImageResource(R.drawable.ic_eat);
        } else if (bean.getType().equals("服装")) {
            viewHolder.ivType.setImageResource(R.drawable.ic_clothes);
        } else if (bean.getType().equals("交通")) {
            viewHolder.ivType.setImageResource(R.drawable.ic_traffic);
        } else if (bean.getType().equals("娱乐")) {
            viewHolder.ivType.setImageResource(R.drawable.ic_play);
        } else {
            viewHolder.ivType.setImageResource(R.drawable.ic_icon);
        }
        viewHolder.tvTitle.setText(bean.getTitle());
        viewHolder.tvExplain.setText(bean.getExplain());
```

```
        viewHolder.tvMoney.setText(String.valueOf(bean.getMoney()));
        viewHolder.tvTime.setText(bean.getTime());
    }
    @Override
    public int getItemCount() {
        return list.size();
    }
    /*内部类*/
    class ViewHolder extends RecyclerView.ViewHolder {
        TextView tvTitle, tvExplain, tvTime, tvMoney;
        ImageView ivType;
        public ViewHolder(View itemView) {
            super(itemView);
            tvTitle = itemView.findViewById(R.id.tv_title_list);
            tvExplain = itemView.findViewById(R.id.tv_explain_list);
            tvTime = itemView.findViewById(R.id.tv_time_list);
            tvMoney = itemView.findViewById(R.id.tv_money_list);
            ivType = itemView.findViewById(R.id.iv_type_list);
        }
    }
}
```

（2）根据分类显示不同列表的要求，创建五个 Fragment 布局文件以及对应的五个 Fragment 类，它们将作为滑屏显示的各个界面元素。由于这五个界面具有相同的布局格式（均为一个 RecyclerView 列表），因此可以只创建一个布局文件，供五个 Fragment 类共用，即创建一个布局文件：account_list.xml，以及五个 Fragment 类：AccountAllFragment、AccountEatFragment、AccountClothesFragment、AccountTrafficFragment、AccountPlayFragment。

account_list.xml 布局文件代码清单如下：

```xml
<?xml version="1.0" encoding="utf-8"?>
<LinearLayout xmlns:android="http://schemas.android.com/apk/res/android"
    android:orientation="vertical" android:layout_width="match_parent"
    android:background="#FFFFFF"
    android:layout_height="match_parent">
    <android.support.v7.widget.RecyclerView
        android:id = "@+id/rv_account"
        android:layout_width="match_parent"
        android:layout_height="match_parent"/>
</LinearLayout>
```

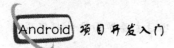

创建的五个Fragment类的代码结构相同,只是类的名称不同,可以在initData()方法中传入不同的测试数据。以AccountAllFragment为例,代码清单如下:

```java
package cc.turbosnail.accountbook;
import android.os.Bundle;
import android.support.v4.app.Fragment;
import android.support.v7.widget.DefaultItemAnimator;
import android.support.v7.widget.DividerItemDecoration;
import android.support.v7.widget.LinearLayoutManager;
import android.support.v7.widget.RecyclerView;
import android.view.LayoutInflater;
import android.view.View;
import android.view.ViewGroup;
import java.util.ArrayList;
import java.util.List;
public class AccountAllFragment extends Fragment {
    private RecyclerView rvAccount;
    private AccountItemAdapter adapter;
    private List<CostBean> list;
    @Override
    public View onCreateView(LayoutInflater inflater, ViewGroup container,
                Bundle savedInstanceState) {
        View view = inflater.inflate(R.layout.account_list, container, false);
        rvAccount = view.findViewById(R.id.rv_account);
        initData();//初始化数据
        adapter = new AccountItemAdapter(this.getContext(), list);
        LinearLayoutManager layout = new LinearLayoutManager(this.getContext());
        rvAccount.setLayoutManager(layout);//设置布局管理器
        rvAccount.setAdapter(adapter);//设置adapter
        rvAccount.setItemAnimator(new DefaultItemAnimator());//设置Item增加、移除动画
        rvAccount.addItemDecoration(new DividerItemDecoration(
                this.getContext(), DividerItemDecoration.VERTICAL));//添加分割线
        return view;
    }
    private void initData() {
        list = new ArrayList<>();
        /*添加一组模拟数据*/
        for (int i = 0; i <5; i++) {
            CostBean bean = new CostBean();
```

```
            bean.setType("用餐");
            bean.setTitle("周末晚餐");
            bean.setExplain("周末同学聚会");
            bean.setTime("2018-11-26");
            bean.setMoney("50");
            list.add(bean);
        }
    }
}
```

（3）修改列表主界面 MainAccountFragment 的布局文件 main_account.xml，添加顶部导航栏和 ViewPager。顶部导航栏通过单选按钮组（RadioGroup、RadioButton）来实现。在本项目中，还添加一个用来进行总金额提醒的文本标签。修改后的 main_account.xml 布局文件代码清单如下：

```
<?xml version="1.0" encoding="utf-8"?>
<android.support.constraint.ConstraintLayout xmlns:android="http://schemas.android.com/apk/res/android"
    xmlns:app="http://schemas.android.com/apk/res-auto"
    android:layout_width="match_parent"
    android:layout_height="match_parent">
    <RadioGroup
        android:id="@+id/rg_select_account"
        android:layout_width="0dp"
        android:layout_height="wrap_content"
        android:orientation="horizontal"
        android:weightSum="5"
        app:layout_constraintLeft_toLeftOf="parent"
        app:layout_constraintRight_toRightOf="parent"
        app:layout_constraintTop_toTopOf="parent">
        <RadioButton
            android:id="@+id/rb_all_account"
            style="@style/TopRadioButtonStyle"
            android:text="全部"/>
        <RadioButton
            android:id="@+id/rb_eat_account"
            style="@style/TopRadioButtonStyle"
            android:text="用餐"/>
        <RadioButton
```

```xml
            android:id="@+id/rb_clothes_account"
            style="@style/TopRadioButtonStyle"
            android:text="服装"/>
        <RadioButton
            android:id="@+id/rb_traffic_account"
            style="@style/TopRadioButtonStyle"
            android:text="交通"/>
        <RadioButton
            android:id="@+id/rb_play_account"
            style="@style/TopRadioButtonStyle"
            android:text="娱乐"/>
    </RadioGroup>
    <android.support.v4.view.ViewPager
        android:id="@+id/vp_list_account"
        android:layout_width="wrap_content"
        android:layout_height="wrap_content"
        app:layout_constraintLeft_toLeftOf="parent"
        app:layout_constraintRight_toRightOf="parent"
        app:layout_constraintTop_toBottomOf="@+id/rg_select_account"/>
</android.support.constraint.ConstraintLayout>
```

由于五个单选按钮具有相同的外观属性,因此可以在项目目录的"values"→"styles.xml"中添加一个自定义样式供五个单选按钮调用,自定义TopRadioButtonStyle样式的代码如下:

```xml
<style name="TopRadioButtonStyle">
    <item name="android:layout_width">0dp</item>
    <item name="android:layout_weight">1</item>
    <item name="android:layout_height">35dp</item>
    <item name="android:button">@null</item>
    <item name="android:gravity">center</item>
    <item name="android:textSize">20sp</item>
    <item name="android:textColor">#999999</item>
```

(4)创建FragmentPagerAdapter适配器对象。自定义构造方法,接收一个Fragment型的List列表,根据列表中的Fragment对象来进行界面适配,代码清单如下:

```java
package cc.turbosnail.accountbook;
import android.support.v4.app.Fragment;
import android.support.v4.app.FragmentManager;
import android.support.v4.app.FragmentPagerAdapter;
```

```java
import java.util.List;
public class AccountFragAdapter extends FragmentPagerAdapter {
    private List<Fragment> fragmentList;//自定义List用来接收多个Fragment对象列表
    /*在构造方法中添加一个 List<Fragment>参数接收数据*/
    public AccountFragAdapter(FragmentManager fm, List<Fragment> fragmentList) {
        super(fm);
        this.fragmentList = fragmentList;
    }
    @Override
    public Fragment getItem(int i) {
        //return null;
        return fragmentList.get(i);//重写,返回自定义List中的元素
    }
    @Override
    public int getCount() {
        //return 0;
        return fragmentList.size();//重写,返回自定义List的大小
    }
}
```

(5)在列表主界面 MainAccountFragment 中创建各个 Fragment 对象,并添加到 ViewPager 中,实现 ViewPager 的切换功能。同时,为顶部导航栏添加事件监听,根据用户选择进行界面切换。MainAccountFragment 代码清单如下:

```java
package cc.turbosnail.accountbook;
import android.os.Bundle;
import android.support.v4.app.Fragment;
import android.support.v4.view.ViewPager;
import android.view.LayoutInflater;
import android.view.View;
import android.view.ViewGroup;
import android.widget.RadioGroup;
import java.util.ArrayList;
import java.util.List;
public class MainAccountFragment extends Fragment {
    private ViewPager viewPager;//实现滑屏
    private RadioGroup rgSelect;//实现顶部导航切换
    public View onCreateView(LayoutInflater inflater, ViewGroup container,
                             Bundle savedInstanceState) {
        View view = inflater.inflate(R.layout.main_account, container, false);
```

```java
            viewPager = view.findViewById(R.id.vp_list_account);
            rgSelect = view.findViewById(R.id.rg_select_account);
            List<Fragment> fragmentList = new ArrayList();
            fragmentList.add(new AccountAllFragment());
            fragmentList.add(new AccountEatFragment());
            fragmentList.add(new AccountClothesFragment());
            fragmentList.add(new AccountTrafficFragment());
            fragmentList.add(new AccountPlayFragment());
        /*当前的Fragment在另一个Fragment中,用getChildFragmentManager()方法*/
            AccountFragAdapter adapter = new AccountFragAdapter(getChildFragmentManager(), fragmentList);
            viewPager.setAdapter(adapter);//添加适配器
            rgSelect.setOnCheckedChangeListener(new RadioGroup.OnCheckedChangeListener() {
                @Override
                public void onCheckedChanged(RadioGroup group, int checkedId) {
                    switch (checkedId) {
                        case R.id.rb_all_account:
                            viewPager.setCurrentItem(0); //viewPager切换到对应索引位置
                            break;
                        case R.id.rb_eat_account:
                            viewPager.setCurrentItem(1);
                            break;
                        case R.id.rb_clothes_account:
                            viewPager.setCurrentItem(2);
                            break;
                        case R.id.rb_traffic_account:
                            viewPager.setCurrentItem(3);
                            break;
                        case R.id.rb_play_account:
                            viewPager.setCurrentItem(4);
                            break;
                        default:
                            break;
                    }
                }
            });
            return view;
        }
    }
```

运行程序，通过滑屏或者点击顶部导航按钮，均可以实现多个Fragment切换显示，运行效果如图4.9所示。

图4.9 带顶部导航按钮的滑屏账目列表

## 4.5 记账界面

**[项目案例9] 为AccountBook项目设计并实现账目添加界面**

记账界面由用户在主界面上点击底部"记一笔"导航按钮时切换打开，它是一个Fragment对象，命名为MainRecordFragment，对应的布局文件是main_record.xml，完善这个布局文件，本例省略布局定位及用于提示的文本标签，对需要进一步响应事件处理的组件定义id属性如下：

```
<?xml version="1.0" encoding="utf-8"?>
<android.support.constraint.ConstraintLayout xmlns:android="http://schemas.android.com/apk/res/android"
    xmlns:app="http://schemas.android.com/apk/res-auto"
    xmlns:tools="http://schemas.android.com/tools"
    android:layout_width="match_parent"
    android:layout_height="match_parent">
    <DatePicker  android:id="@+id/dp_time" />  <!--日期选择器-->
    <EditText  android:id="@+id/edt_title"
        android:hint="输入消费的主题"/>
    <Spinner  android:id="@+id/sp_type"
        android:entries="@array/cost"/>  <!--下拉列表项:用餐、服装、交通、娱乐-->
```

```xml
<EditText  android:id="@+id/edt_money"
    android:hint="输入金额"/>
<EditText  android:id="@+id/edt_explain"
    android:hint="对消费的详细说明"/>
<ImageView  android:id="@+id/iv_camera"/>  <!--用来打开相机拍照-->
<ImageView  android:id="@+id/iv_photo"/>   <!--用来选择图片库中已有的照片-->
<ImageView  android:id="@+id/iv_image"
    android:src="@drawable/ic_small_logo"/>  <!--用来预览选择的图片-->
<Button  android:id="@+id/btn_add_record"
    android:text="确认"/>
</android.support.constraint.ConstraintLayout>
```

调整布局,界面效果如图4.10所示。

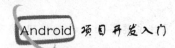

图4.10  记账界面

## 4.6  图表统计界面

图表统计是当前移动端应用开发的一个重要功能,对数据通过折线图、雷达图等方式进行统计显示,既能通过形象、直观的视图大幅度地提升用户体验感,又能对数据进行各种专业分析以提升软件的应用价值。Android提供了基于Canvas的基础绘图框架允许用户进行图形图像绘制,但是一些开源的图表插件可以为开发者提供更友好、更专业的绘图功能,在应用开发中可以使用这些优秀的开源项目成果,以减少重复劳动,提高开发效率。

MPAndroidChart是一个强大、易用的图表插件库,它提供了条型图、折线图、饼图、散点图、泡泡图、直方图、雷达图、组合图等多种图表,而且可以实现缩放、拖动等多种动画效果,以及事件响应等操作,其在 Github 上的下载地址为 https://github.com/PhilJay/MPAndroidChart,开发文档及示例代码均可在该地址上查看。

### 4.6.1 在 Android Studio 中添加第三方库(依赖)

(1) 添加 jcenter 中的依赖库。

在使用 RecylerView 时,首先通过在项目目录"Gradle Scripts"→"build.gradle(Module:app)"中添加"implementation 'com.android.support:recyclerview-v7:28.0.0'"语句导入依赖库,然后才能使用这个组件。Android Studio 在添加外部依赖库时,会首先在本地项目中查找,如果存在,则直接使用,如 RecylerView 库就是在本地已经存在,可以直接导入使用。如果不存在,则会到默认的代码仓库 jcenter 中去下载,此时需要联网。如添加"implementation 'com.squareup.okhttp:okhttp:2.0.0'"库就需要联网下载。查找和添加 jcenter 中的依赖库,既可以手动在 build.gradle(Module:app)文件中添加,也可以在"File"→"Project Structure"窗口中选择或者输入依赖库名称并查找后直接添加,示意图如图 4.11 所示。

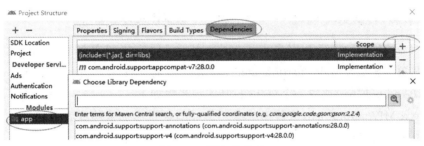

图 4.11 添加依赖库

(2) 添加 maven 中的依赖库。

Android Studio 也支持 maven 中的依赖库。首先在"Gradle Scripts"→"build.gradle (Project:项目名)"中添加如下代码:

```
allprojects {
    repositories {
        maven { url 'https://jitpack.io' }
    }
}
```

然后在"Gradle Scripts"→"build.gradle(Module:app)"中添加相关依赖,如本项目中要使用的 MPAndroidChart 依赖:

implementation 'com.github.PhilJay:MPAndroidChart:v3.0.3'

此时同样需要根据提示"sync"进行联网同步,可以在控制台看到相关下载信息。下载成功后,添加的依赖库可以在项目中通过导入包的形式使用。

(3) 添加 jar 包。

在项目开发的过程中,还可以先下载依赖库的 jar 包,将其直接添加到项目中使用。以 MPAndroidChart 为例:首先在 Github 上下载 MPAndroidChart-3.0.3.jar 文件,然后在项目中切换到"project"视图,把文件复制到libs目录中,如图4.12所示。

最后,在导入包上点击右键,选择"Add As Library",如图4.13所示。等待同步完成后,切换项目视图到"Android"模式下,打开"Gradle Scripts"→"build.gradle(Module:app)"可以看到,依赖中添加了如下语句:

implementation files('libs/MPAndroidChart-v3.0.3.jar')

图4.12　导入jar包　　　　　　图4.13　jar包添加为依赖库

### 4.6.2　在项目中添加图表——基于MPAndroidChart

**[项目案例10] 为AccountBook项目实现一个饼状图统计界面**

本项目图表统计界面由用户点击"图表统计"导航按钮时切换打开,它也是一个Fragment对象,命名为MainChartFragment,对应的布局文件是main_chart.xml。在该布局中,基于MPAndroidChart放置一个饼状图,其实现步骤如下:

(1) 导入MPAndroidChart依赖库。参照上一节通过添加依赖或者直接导入jar包的形式在项目中添加依赖库。

(2) 在main_chart.xml布局文件中添加一个饼状图。

```xml
<?xml version="1.0" encoding="utf-8"?>
<android.support.constraint.ConstraintLayout xmlns:android="http://schemas.android.com/apk/res/android"
    xmlns:app="http://schemas.android.com/apk/res-auto"
    android:layout_width="match_parent"
    android:layout_height="match_parent">
    <com.github.mikephil.charting.charts.PieChart
        android:id="@+id/chart_account"
        android:layout_width="match_parent"
        android:layout_height="match_parent"
        app:layout_constraintBottom_toBottomOf="parent"
        app:layout_constraintLeft_toLeftOf="parent"
        app:layout_constraintRight_toRightOf="parent"
        app:layout_constraintTop_toTopOf="parent"/>
</android.support.constraint.ConstraintLayout>
```

（3）在 MainChartFragment 中对饼状图进行初始化设置。主要包括一些基本属性设置，以及用两个 ArrayList 对象分别定义对应的显示数据和颜色。使用 PieDataSet 添加数据列表，并为它设置颜色、字体等属性，然后将其添加给 PieData，为 PieData 设置显示外观，最后为饼状图设置构建好的 pieData 对象为数据源。

饼状图还有很多属性可以实现灵活多样的外观效果、动画效果，并可以实现添加用户事件响应等操作，具体可参阅其 API 文档。

```java
package cc.turbosnail.accountbook;
import android.graphics.Color;
import android.os.Bundle;
import android.support.v4.app.Fragment;
import android.view.LayoutInflater;
import android.view.View;
import android.view.ViewGroup;
import com.github.mikephil.charting.charts.PieChart;
import com.github.mikephil.charting.components.Description;
import com.github.mikephil.charting.data.PieData;
import com.github.mikephil.charting.data.PieDataSet;
import com.github.mikephil.charting.data.PieEntry;
import com.github.mikephil.charting.formatter.PercentFormatter;
import java.util.ArrayList;
public class MainChartFragment extends Fragment {
    private PieChart chartAccount;
    public View onCreateView(LayoutInflater inflater, ViewGroup container,
                    Bundle savedInstanceState) {
        View view = inflater.inflate(R.layout.main_chart, container, false);
        chartAccount = view.findViewById(R.id.chart_account);
        chartAccount.setDrawHoleEnabled(false);//设置为实心
        Description description = new Description();
        description.setText("消费分类统计图");
        chartAccount.setDescription(description);//设置右下方文本标签内容
        ArrayList<PieEntry> entries = new ArrayList<>();//数据列表
        entries.add(new PieEntry(40.0f, "用餐"));//(数据,分类名)
        entries.add(new PieEntry(27.0f, "服装"));
        entries.add(new PieEntry(10.0f, "交通"));
        entries.add(new PieEntry(23.0f, "娱乐"));
        PieDataSet dataSet = new PieDataSet(entries, "");//构建 PieDataSet
        dataSet.setValueTextSize(18f);//设置字体大小
        ArrayList<Integer> colors = new ArrayList<>();//定义颜色列表
```

```
        colors.add(Color.rgb(205, 205, 205));
        colors.add(Color.rgb(114, 188, 223));
        colors.add(Color.rgb(255, 123, 124));
        colors.add(Color.rgb(57, 135, 200));
        dataSet.setColors(colors);//为 PieDataSet 设置颜色
        PieData pieData = new PieData(dataSet);//构建 PieData
        pieData.setValueFormatter(new PercentFormatter());//按百分比显示,默认为直接显示
数值
        chartAccount.setData(pieData);//为图表添加 PieData 数据
        chartAccount.invalidate();//刷新
        return view;
    }
}
```

运行程序,点击"图表分析"导航按钮,可以看到运行效果如图4.14所示。

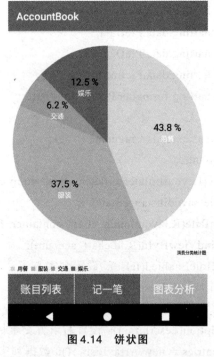

图4.14　饼状图

Android作为一个庞大的生态系统,有大量优秀的开源项目成果,涵盖了界面开发、数据处理、网络访问、信息安全等各个方面,建议读者在学习Android开发的过程中主动关注,并加以应用。

# 第5章 项目中的数据存取

数据的存储和读取是应用软件开发的一个重要问题,一些数据需要持久化保存,比如用户的账号信息、用户提交保存的数据、允许用户访问的数据等,这些数据可以存储在本地设备上,也可以存储在远端的服务器上,需要访问时程序可以对这些数据进行读取。当数据需要在本地存取时,Android提供了Shared Preferences、File、SQLite等方式供开发者选用。

## 5.1 引导页不再出现——Shared Preferences

引导页在程序安装后第一次运行时出现,当用户再次运行程序时,引导页就不需要再出现,因此需要在程序中设置一个持久的标记值来表示程序是否运行过,当程序运行过一次后,该标记的值发生变化,程序再次启动时,根据变化后的标记值,让程序不再跳转到欢迎页,而是跳转到登录界面或者主界面上去。这个持久化的标记可以用Shared Preferences实现。

### 5.1.1 Shared Preferences

Shared Preferences是Android提供的一个轻量级的数据存储工具类,用来保存一些简单数据和配置信息,例如记住用户名和密码、记住用户设置的音效等配置信息、保存一些程序状态值等。Shared Preferences中存储的数据会记录在程序的"data/data/<packagename>/shared_prefs"目录下,以XML文件的形式存储,所以数据一旦存储,就可以反复被访问。

Shared Preferences通过<Key,Value>键值对的结构存储数据,Key作为Value的访问标识符由开发者自己定义,一个应用程序中Key的名字是唯一的。Value主要支持一些简单的数据类型:int、long、Float、boolean、String、Set、Map等。Shared Preferences的基本用法如下:

(1)存储数据。通过当前上下文环境(Context)的getSharedPreferences(String name, int mode)方法获得SharedPreferences对象,该方法的第一个参数是保存的XML文件名称,如果该名称已经存在,则打开;如果不存在,则新建。第二个参数是访问权限,可以使用常量:MODE_PRIVATE、MODE_WORLD_READABLE、MODE_WORLD_WRITEABLE。

接着获得SharedPreferences的Editor对象,通过该对象的put方法保存数据,然后通过commit方法进行提交。Editor对象可以理解为一个编辑器,用来进行数据操作,Editor对象

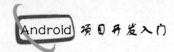

的主要方法列表如表5.1所示。

例如,创建/打开一个名称为"data"的SharedPreferences对象,存入两组数据的步骤如下:

表5.1  Editor方法列表

| 序号 | 方法名称 | 方法说明 |
|---|---|---|
| 1 | putBoolean(String key, boolean value) | 以<key, value>的形式存入一个value值,key作为标识符在读取数据时使用,value的数据类型应该和方法对应 |
| 2 | putFloat(String key, float value) | |
| 3 | putInt(String key, int value) | |
| 4 | putLong(String key, long value) | |
| 5 | putString(String key, String value) | |
| 6 | putStringSet(String key, Set<String> values) | |
| 7 | remove(String key) | 移除key对应的数据 |
| 8 | clear() | 清空SharedPreferences中的所有数据 |
| 9 | boolean commit() | 提交操作,修改数据的操作必须通过该方法才能生效 |

```
SharedPreferences sp = getSharedPreferences("data",MODE_PRIVATE);
Editor editor = sp.edit();
editor.putString("name", "张三");
editor.putInt("age", 8);
editor.commit();
```

(2)取出数据。需要取出数据时,同样先通过当前上下文环境(Context)的getSharedPreferences(String name, int mode)方法获取SharedPreferences对象,然后通过对应数据类型的get方法取值,如boolean getBoolean(String key, boolean defValue)、int getInt(String key, int defValue)等,第一个参数为所存储数据的Key,通过它返回数据,第二个参数是默认值,即当没有这个Key对应数据时,默认返回的值。示例代码如下:

```
SharedPreferences sp = getSharedPreferences("data",MODE_PRIVATE);
String name = sp.getString("name",null);
Float score = sp.getFloat ("score","0.0f");//没有这个值,则返回0.0f
```

(3)编辑数据。编辑数据主要是用Editor对象的remove(String key)和clear()方法,分别用来删除一个Key对应的数据和清空当前SharedPreferences的所有数据。注意:和添加数据一样,完成操作后必须要通过Editor对象的commit()方法提交才能完成修改。示例代码如下:

```
SharedPreferences sp = getSharedPreferences ("data",MODE_PRIVATE);
Editor editor = sp.edit();
```

```
editor.remove("name");
editor.clear();
editor.commit();
```

### 5.1.2 让引导页只运行一次

让引导页只运行一次的设计思想为：当 Activity 第一次运行时创建一个 SharedPreferences 对象，存储一个 boolaen 型的标记变量，并将它赋值为 true。然后读取程序并检查这个标记变量，如果值为 true（表示是第一次运行程序），则修改为 false，并跳转到引导页；如果值为 false，则跳转到主界面，引导页不再出现。

[项目案例 11] 让 AccountBook 项目的引导页只执行一次

当程序第一次运行时，从欢迎页先跳转到滑屏引导页；当程序再次运行时，从欢迎页直接跳转到登录页，不再进入滑屏引导页。

（1）首先创建登录 Activity。新建一个"Empty Activity"，命名为 LoginActivity，由于项目案例中已经创建了登录界面的布局文件 activity_login.xml，所以在创建 Activity 时取消掉 "Generate Layout File" 的勾选，然后在生成的 LoginActivity 中手动添加 "setContentView（R.layout.activity_login）;"语句，如图 5.1 所示。

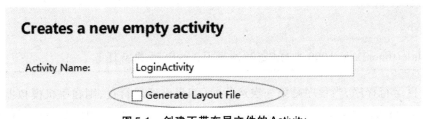

图 5.1 创建不带布局文件的 Activity

```
protected void onCreate(Bundle savedInstanceState) {
    setContentView(R.layout.activity_login);//手动添加该句
    super.onCreate(savedInstanceState);
}
```

（2）修改 GuideActivity 中的跳转语句，让它跳转到登录界面，示例代码如下：
Intent intent = new Intent(GuideActivity.this, LoginActivity.class);
（3）修改 WelcomeActivity.java 代码中的计时器线程，根据是否第一次运行程序跳转到不同界面上去，代码清单如下：

```
Timer timer = new Timer();//创建定时器
/*实现 TimerTask 接口,任务在 run 方法中*/
TimerTask timerTask = new TimerTask() {
    @Override
```

```java
public void run() {
    SharedPreferences sp = getSharedPreferences("use_status", MODE_PRIVATE);
    boolean isFirst = sp.getBoolean("isFirst",true);//isFirst不存在,则默认返回true
    if (!isFirst) {
        /*不是第一次启动,则跳转到登录页*/
        Intent intent = new Intent(WelcomeActivity.this,LoginActivity.class);
        startActivity(intent);
        WelcomeActivity.this.finish();
    } else {
        /*第一次启动,向sp中添加一个标识变量isFirst,值为false*/
        SharedPreferences.Editor editor = sp.edit();
        editor.putBoolean("isFirst", false);
        editor.commit();
        /*第一次启动,跳转到欢迎页*/
        Intent intent = new Intent(WelcomeActivity.this, GuideActivity.class);
        startActivity(intent);
        WelcomeActivity.this.finish();
    }
  }
};
timer.schedule(timerTask, 1000 * 3);//定时器延迟3秒启动,执行任务
```

此时,再运行程序,当程序是第一次运行时(如果已经运行过,则在手机模拟器桌面上卸载掉AccountBook),从欢迎界面会跳转到引导页,当程序再次运行时,欢迎页结束后会自动跳转到登录界面,引导页不再出现。

### 5.1.3 记住用户名和密码

在登录界面上显示记住的用户名和密码是应用开发中的一个常见功能,通过SharedPreferences可以比较简单地实现这一功能,其设计思想为:登录界面启动时,先读取SharedPreferences中存储的用户名和密码,将其赋值到对应的编辑框里。用户点击登录按钮时,验证用户名和密码,如果验证通过,则将用户名和密码存储在SharedPreferences中。

**[项目案例12]** 在AccountBook项目的登录界面上实现记住用户名和密码的功能

(1) 在LoginActivity初始化时,首先读取SharedPreferences中的值,记录用户名和密码的状态变量isCheck,如果值为true,则读取username和password的值,并将其赋给界面上的用户名和密码框,同时设置复选框为选中状态。代码实现如下:

```
sp = getSharedPreferences("user", MODE_PRIVATE);
if (sp.getBoolean("isRemeber", false)) {
    edtUsername.setText(sp.getString("username", ""));
    edtPassword.setText(sp.getString("password", ""));
    cbRemeber.setChecked(true);
}
```

（2）在登录按钮事件响应过程中，验证用户名和密码正确后，先将当前的用户名、密码存储到 SharedPreferences 里面，并保存用户是否记住用户名和密码的状态值，再进行界面跳转。

```
btnLogin.setOnClickListener(new View.OnClickListener() {
    @Override
    public void onClick(View v) {
        String name = edtUsername.getText().toString();
        String password = edtPassword.getText().toString();
        if (name.equals("张三") && password.equals("123")) { //此处用一组固定值测试
            editor = sp.edit();
            editor.putBoolean("isRemeber", cbRemeber.isChecked());
            editor.putString("username", name);
            editor.putString("password", password);
            editor.commit();
            Intent intent = new Intent(LoginActivity.this, MainActivity.class);
            startActivity(intent);
            LoginActivity.this.finish();
        } else {
            Toast.makeText(LoginActivity.this, "用户名或密码错误!", Toast.LENGTH_SHORT).
            show();
        }
    }
});
```

本例省略了变量声明及初始化的相关操作，请读者自行完善。

要保证 SharedPreferences 的名称统一，这样才是对同一个 SharedPreferences 进行操作，例如本例中的"user"。同时，存储和读取时的 key 值也要对应一致，如本例中的"username""password"。

## 5.2 本地数据存储——SQLite数据库

### 5.2.1 SQLite基本用法

SQLite是一款轻量级、嵌入式的关系型数据库。SQLite嵌入使用它的应用程序中,它们共用相同的进程空间,从外部看,它并不像一个独立的RDBMS(Relational Database Management System),但在进程内部却是完整的,自包含数据库引擎。SQLite具有开源、轻量级、独立性、支持事务、跨平台、多语言支持、占内存小、安全可靠等众多优点,而且支持大多数SQL92标准的查询语句,这些特性使得SQLite被大量地用于资源受限的嵌入式设备中。

Android系统集成了SQLite数据库,通过提供SQLiteDatabase、SQLiteOpenHelper来管理和操作数据库。其基本使用流程如下:

(1) 构建SQLiteOpenHelper的子类。

SQLiteOpenHelper顾名思义是一个数据库管理的帮助类,可以把创建数据库、创建数据表、数据库版本管理的操作放在这里,由它来保证当这个数据库已经存在时这些创建语句不会被重复执行、数据库版本更新等一些涉及数据库管理的业务逻辑。

SQLiteOpenHelper的子类,至少需要实现以下三种方法:

- 构造方法 super(Context context, String name, CursorFactory cursorFactory, int version)。调用父类SQLiteOpenHelper的构造方法。这个方法需要四个参数:上下文环境(例如,一个Activity)、数据库名字、一个可选的游标工厂(通常是null)、一个表示数据库版本的整数。
- public void onCreate(SQLiteDatabase db)。创建数据库时会调用,所以创建数据库和表的代码可以放在这个方法里。注意:该方法只会调用一次,如果数据库已经存在,那么打开数据库时是不会再次调用这个方法的。
- public void onUpgrade(SQLiteDatabase db, int oldVersion, int newVersion)。它需要三个参数:一个SQLiteDatabase对象、一个旧的版本号和一个新的版本号。该方法在数据库需要升级时才会调用。数据库字段发生变化时,在该方法中添加修改数据表的代码,如果传入的新版本号大于已有版本号,则该方法会被调用执行。

一个典型的SQLiteOpenHelper子类定义如下:

```
import android.content.Context;
import android.database.sqlite.SQLiteDatabase;
import android.database.sqlite.SQLiteOpenHelper;
public class DBHelper extends SQLiteOpenHelper {
    private static final String DATABASE_NAME = "account.db";//数据库名称
    private static final int DATABASE_VERSION = 1;//数据库版本号
    /*创建数据表的SQL语句字符串*/
```

```java
    private String userCreate = "CREATE TABLE IF not exists user(" +
        "user_id INTEGER PRIMARY KEY AUTOINCREMENT," +
        "user_name TEXT," +
        "user_password TEXT)";
    public DBHelper(Context context, String name,SQLiteDatabase.CursorFactory factory, int version) {
        // super(context, name, factory, version);
        super(context, DATABASE_NAME, null, DATABASE_VERSION);//构造方法
    }
    @Override
    public void onCreate(SQLiteDatabase db) {
        db.execSQL(userCreate);//创建表
    }
    @Override
    public void onUpgrade(SQLiteDatabase db, int oldVersion, int newVersion) {
        db.execSQL("ALTER TABLE user ADD COLUMN user_tel TEXT");//版本号更新后才会执行
    }
}
```

SQLite 支持五种数据类型:NULL、INTEGER(整数)、REAL(浮点数)、TEXT(字符串文本)、BLOB(二进制数),也接受 varchar(n)、char(n)、decimal(p,s)等类型,在运算或保存时会自动转换成五种数据类型之一。SQLite 支持弱类型(Manifest Typing),可以将各种类型数据存到同一字段中,即允许保存数据类型不一致的数据到对应字段中,如 INTEGER 中可以存放 TEXT 类型数据。

(2) 构建 SQLiteDatabase 对象。

Android 将数据库访问相关的操作封装成 SQLiteDatabase 类,通过这个类的对象来进行各种数据操作。我们可以通过 SQLiteOpenHelper 类的 getWritableDatabase()和 getReadableDatabase()方法得到一个 SQLiteDatabase 对象。

示例代码如下:

```java
DBHelper helper = new DBHelper(MainActivity.this,"account.db");
SQLiteDatabase db = helper.getWritableDatabase();
```

此时,如果数据库 account 已经存在,则打开这个数据库;如果不存在,则执行 DBHelper 中的 onCreate 方法,新建这个数据库。注意:getReadableDatabase()并不是以只读方式打开数据库的,而是先执行 getWritableDatabase(),调用失败的情况下才会以只读方式打开数据库。

(3) 数据操作。

① 增、删、改操作。SQLiteDatabase 通过 public void execSQL(String sql)方法可以执行 SQL 的新建表、增删改语句功能,开发者构建好 SQL 语句,作为参数传递给该方法执行,常

用的简单SQL语句示例如下：

String sql ="INSERT INTO user (user_name, user_password) VALUES('张三', '666')";//插入一条记录
String sql = "DELETE FROM user WHERE user_name ='张三'";//删除一条记录
String sql = "SELECT * FROM user WHERE user_name ='张三'";//查询所有记录
String sql = "UPDATE user SET user_password ='aaa' WHERE user_id=3";//修改一条记录

　　SQLiteDatabase还支持带占位符的SQL语句，通过public void execSQL（String sql，Object[] bindArgs）方法执行。该方法的第一个参数为SQL语句，第二个参数为占位符数组，参数值在数组中的顺序要和占位符的位置对应。示例如下：

String sql = "insert into user（user_name，user_password）values(?,?)";
db.execSQL(sql,new String[]{"李四", "333"});

　　② 查询操作。SQLiteDatabase通过public Cursor rawQuery（String sql，String[] selectionArgs)方法可以执行SQL查询语句，返回Cursor型的结果集，以便开发者进一步处理查询结果。selectionArgs是占位符，如果不使用的话，可以为null。结果集以二维表的结构组成，初始状态下，游标指针指向第一行的上方，在遍历数据时通过游标移动确定行位置，通过对应字段的索引位置确定列位置，即可得到一个具体的数据值。示意图如图5.2所示。

| user_id | user_name | user_password |
|---|---|---|
| 1 | 张三 | 666 |
| 2 | 李四 | 333 |
| 3 | 王五 | 555 |

图5.2　结果集遍历示意图

　　Android对Cursor型数据的处理提供了大量方法可以使用：

- boolean isFirst()、isLast() 是否第一条、最后一条记录。
- boolean isBeforeFirst() 是否指向第一条记录之前。
- boolean isAfterLast() 是否指向最后一条记录之后。
- boolean moveToFirst() 移动到第一条记录。
- boolean moveToLast() 移动到最后一条记录。
- boolean moveToNext() 移动到下一条记录。
- boolean move(int offset) 移动到offset索引位置的记录。
- boolean moveToPrevious() 移动到上一条记录。
- String[] getColumnNames () 以数组形式返回结果集中所有字段名称(即列名)。
- String getString（int columnIndex）、int getInt（int columnIndex）、long getLong（int columnIndex）、float getFloat（int columnIndex）、double getDouble（int columnIndex）等方法，返回结果集对应索引位置的数据值，columnIndex为索引列的编号，从0开始。

- int getColumnIndex（String columnName）返回结果集中字段名对应的列索引号。
- boolean isNull(int columnIndex) 判断指定索引位置的数据值是否为空。
- int getCount() 返回结果集中的记录总数。
- int getPosition() 返回当前游标所指向的行数。
- close() 释放游标资源。

一个典型的简单处理过程如下：

```
Cursor cursor=db.rawQuery("SELECT * FROM user",null);
while (cursor.moveToNext()) {
    int id = cursor.getInt(0);//获取第一列对应的值
    String name = cursor.getString(1);
    int password = cursor.getString (2);
}
cursor.close();
```

### 5.2.2 项目案例中的数据库应用

#### 1. 数据库管理类

**[项目案例13] 在AccountBook项目中创建数据库管理类**

根据本项目的数据存储需求，可以创建一个数据库，包含两张数据表。一张用户表用来存储用户信息：用户id、用户名、用户密码。当注册新用户时，保存注册信息；当用户登录时，在这张表中查询是否存在该用户。另一张账目表用来存储账目信息：账目id、账目类型、账目标题、账目描述、账目金额、账目图片、用户id。两张表中的用户id保持一致，用来标识账目对应的用户。当用户登录后，可以在账目列表中看到自己的账目信息，可以在添加账目界面添加新的账目，也可以通过长按列表项删除列表中的一条账目。

数据库名定义为account_db，数据库版本号定义为1，数据表名分别为user、cost，在AccountBook中新建一个Java Class，命名为DBHelper。代码清单如下：

```
package cc.turbosnail.accountbook;
import android.content.Context;
import android.database.sqlite.SQLiteDatabase;
import android.database.sqlite.SQLiteOpenHelper;
public class DBHelper extends SQLiteOpenHelper {
    private static final String DATABASE_NAME = "account.db";//数据库名称
    private static final int DATABASE_VERSION = 1;//数据库版本号
    /*创建用户表user的SQL语句*/
    private String userCreate = "CREATE TABLE IF not exists user(" +
```

```
        "user_id INTEGER PRIMARY KEY AUTOINCREMENT," +
        "user_name TEXT," +
        "user_password TEXT)";
/*创建账目表cost的SQL语句*/
private String costCreate = "CREATE TABLE IF NOT EXISTS cost(" +
        "cost_id INTEGER PRIMARY KEY AUTOINCREMENT," +
        "cost_type TEXT," +
        "cost_money REAL," +
        "cost_title TEXT," +
        "cost_explain TEXT," +
        "cost_time TEXT," +
        "cost_pic TEXT," +
        "user_id INTEGER)";
public DBHelper(Context context) {
    super(context, DATABASE_NAME, null, DATABASE_VERSION);//构造方法
}
@Override
public void onCreate(SQLiteDatabase db) {
    db.execSQL(userCreate);//创建表
    db.execSQL(costCreate);//创建表
}
@Override
public void onUpgrade(SQLiteDatabase db, int oldVersion, int newVersion) {
}
}
```

### 2. 用户注册、登录功能实现

**[项目案例14] 在AccountBook项目中实现用户注册功能、用户登录功能**

（1）创建用户实体类。用户注册和登录涉及的数据表是user表，注册时用户信息存储到表中，登录时根据用户输入信息到该表中查询对应用户信息是否存在。首先根据数据表的设计及业务逻辑需要，在项目中新建一个"Java Class"作为用户实体类，类名为UserBean，声明与数据表对应的三个成员变量，并生成set/get访问器方法。代码清单如下：

```
package cc.turbosnail.accountbook;
public class UserBean {
    private int userID;
    private String userName;
```

```java
    private String passwrod;
    public int getUserID() {
        return userID;
    }
    public void setUserID(int userID) {
        this.userID = userID;
    }
    public String getUserName() {
        return userName;
    }
    public void setUserName(String userName) {
        this.userName = userName;
    }
    public String getPasswrod() {
        return passwrod;
    }
    public void setPasswrod(String passwrod) {
        this.passwrod = passwrod;
    }
}
```

（2）创建用户数据操作类。在项目中新建一个"Java Class"，类名为 UserDao，在该类中封装对用户信息数据操作。通过 DBHelper 类获取到 SQLiteDatabase 数据库操作对象，然后根据项目需求封装数据库操作的业务逻辑代码，在本例中封装了四个方法，分别用来进行添加用户、查询用户名和密码是否存在、获取用户 id、关闭数据库连接。代码清单如下：

```java
package cc.turbosnail.accountbook;
import android.content.Context;
import android.database.Cursor;
import android.database.sqlite.SQLiteDatabase;
public class UserDAO {
    private DBHelper dbHelper;
    private SQLiteDatabase db;
    public UserDAO(Context context) {
        dbHelper = new DBHelper(context);
        db = dbHelper.getWritableDatabase();
    }
    /*添加用户时首先判断该用户名是否已经存在,如果存在,返回false,不进行数据插入操作*/
```

```java
public boolean addUser(UserBean user) {
    boolean flag;
    String sql = "SELECT * FROM user WHERE user_name=?";
    Cursor cursor = db.rawQuery(sql, new String[]{user.getUserName()});
    if (cursor != null && cursor.getCount() > 0) {
        flag = false;//如果用户名已经存在,则返回false
    } else {
        String sqlIn = "INSERT INTO user(user_name,user_password) VALUES(?,?)";
        db.execSQL(sqlIn, new String[]{user.getUserName(), user.getPasswrod()});
        flag = true;//插入数据,并返回true
    }
    cursor.close();
    return flag;
}
/*检查用户名和密码,如果正确则返回true*/
public boolean checkIn(String userName, String password) {
    boolean isAllow;
    String sql = "SELECT * FROM user WHERE user_name=? AND user_password=?";
    Cursor cursor = db.rawQuery(sql, new String[]{userName, password});
    if (cursor != null && cursor.getCount() > 0) {
        isAllow = true;
    } else {
        isAllow = false;
    }
    cursor.close();
    return isAllow;
}
public Integer getID(String username) {
    Integer userID = 0;
    String sql = "SELECT user_id FROM user WHERE user_name=?";
    Cursor cursor = db.rawQuery(sql, new String[]{username});
    if (cursor.moveToNext()) {
        userID = cursor.getInt(0);
    }
    return userID;
}
public void close() {
    if (db != null) {
        db.close();
```

```
        }
    }
}
```

(3) 创建注册 Activity。新建一个 "Empty Activity"，命名为 RegisterActivity，由于项目案例中已经创建了注册界面的布局文件 activity_register.xml，因此在创建 Activity 时取消掉 "Generate Layout File" 的勾选，然后在生成的 RegisterActivity 中手动添加 "setContentView（R.layout.activity_register）;" 语句。

为登录界面上的 "注册新用户" 文本标签添加事件响应，用户点击时，跳转到注册界面上去。在 LoginActivity 代码中获取文本标签 tv_register_login 的对象，并添加监视器代码如下：

```
tvRegister.setOnClickListener(new View.OnClickListener() {
    @Override
    public void onClick(View v) {
        Intent intent = new Intent(LoginActivity.this, RegisterActivity.class);
        startActivity(intent);
    }
});
```

(4) 在 RegisterActivity 中添加注册功能代码。首先检查用户输入的用户名是否为空，两次密码是否一致，然后构建 UserBean 对象，作为参数传递给 UserDAO 对象的 addUser 方法，该方法会返回是否存在该用户或者数据添加成功的 boolean 型标识。

```java
package cc.turbosnail.accountbook;
import android.content.Intent;
import android.os.Bundle;
import android.support.v7.app.AppCompatActivity;
import android.view.View;
import android.widget.Button;
import android.widget.EditText;
import android.widget.Toast;
public class RegisterActivity extends AppCompatActivity {
    private EditText edtUsername, edtPassword, edtRepeat;
    private Button btnRegister;
    private UserBean userBean;
    @Override
    protected void onCreate(Bundle savedInstanceState) {
        super.onCreate(savedInstanceState);
        setContentView(R.layout.activity_register);
        initView();
```

```java
        userBean = new UserBean();
        btnRegister.setOnClickListener(new View.OnClickListener() {
            @Override
            public void onClick(View v) {
                String name =edtUsername.getText().toString().trim();
                String password =edtPassword.getText().toString().trim();
                String repeat = edtRepeat.getText().toString().trim();
                if(name==null||name.equals("")){
                    Toast. makeText(RegisterActivity. this, "用户名不能为空,请输入", Toast. LENGTH_SHORT).show();
                }else{
                    if (!password.equals(repeat)) {
                        Toast.makeText(RegisterActivity.this, "两次输入的的密码不一致,请重新输入", Toast.LENGTH_SHORT).show();
                    } else {
                        userBean.setUserName(name);
                        userBean.setPasswrod(password);
                        if (new UserDAO(RegisterActivity.this).addUser(userBean)) {
                            Intent intent = new Intent(RegisterActivity.this, LoginActivity.class);
                            startActivity(intent);
                            RegisterActivity.this.finish();
                        } else {
                            Toast. makeText(RegisterActivity. this, "用户名已存在!", Toast. LENGTH_SHORT).show();
                        }
                    }
                }
            }
        });
    }
    public void initView() {
        edtUsername = findViewById(R.id.edt_username_register);
        edtPassword = findViewById(R.id.edt_password_register);
        edtRepeat = findViewById(R.id.edt_repeat_register);
        btnRegister =  findViewById(R.id.btn_register);
    }
}
```

（5）在LoginActivity中添加登录功能代码。此时,不再是使用一组固定值来进行验证,而是根据用户输入的用户名和密码,到数据库中查找是否存在该用户,如果验证通过则将用户 id 也存入 SharedPreferences 中。代码清单如下:

```java
btnLogin.setOnClickListener(new View.OnClickListener() {
    @Override
    public void onClick(View v) {
        String name = edtUsername.getText().toString().trim();
        String password = edtPassword.getText().toString().trim();
        if (name == null || name.equals("")) {
            Toast. makeText(LoginActivity. this, " 用 户 名 不 能 为 空 , 请 输 入 ", Toast.LENGTH_SHORT).show();
        } else {
            //调用 UserDAO 中的 checkIn 方法,检查登录
            if (userDAO.checkIn(name, password)) {
                editor = sp.edit();
                int userID = userDAO.getID(name);//调用 UserDAO 中 getID 方法,获取用户 id
                editor.putInt("userID", userID);
                editor.putString("isCheck", String.valueOf(cbRemeber.isChecked()));
                editor.putString("username", userBean.getUserName());
                editor.putString("password", userBean.getPasswrod());
                editor.commit();
                Intent intent = new Intent(LoginActivity.this, MainActivity.class);
                startActivity(intent);
                LoginActivity.this.finish();
            } else {
                Toast. makeText(LoginActivity. this, " 用 户 名 或 密 码 错 误 ！ ", Toast.LENGTH_SHORT).show();
            }
        }
    }
});
```

运行程序,在程序进入登录界面后,通过点击"注册新用户"跳转到注册界面,用户可以在这里注册一个账号,注册成功后跳转到登录界面,此时,可以通过注册的账号进行登录。如注册多个账号,则每个账号均可进行登录,如图 5.3 所示。

图5.3 账号登录

### 3. 账目信息添加与数据刷新

**[项目案例15]** 在AccountBook项目中实现账目信息添加功能

（1）创建账目实体类。账目信息存储在coat表中，添加账目即在表中添加一组数据，显示账目列表即查询用户的所有账目信息作为列表数据源，删除账目即从表中删除一组数据。根据数据表的设计及业务逻辑需要，在项目中新建一个"Java Class"作为账目信息的实体类，类名为CostBean，声明与数据表对应的成员变量，并生成set/get访问器方法。代码清单如下：

```
package cc.turbosnail.accountbook;
public class CostBean {
    private int costID;//账目id
    private int userID;//用户id
    private float money;//金额
    private String type;//账目类型
    private String title;//账目内容
    private String explain;//账目详细描述
    private String time;//时间
    private String picPath;//图片路径
    public void setCostID(int costID) {
        this.costID = costID;
    }
    public int getCostID() {
        return costID;
    }
……此处省略其他成员变量的set/get方法。
}
```

（2）创建账目数据操作类。在项目中新建一个"Java Class"，类名为CostDao，在该类中封装对账目信息的相关操作。通过DBHelper类获取到SQLiteDatabase数据库的操作对象，然后根据项目需求封装数据库操作的业务逻辑代码，在本案例中封装了四个方法，分别用来进行账目添加、查询、删除和关闭数据库连接。代码清单如下：

```java
package cc.turbosnail.accountbook;
import android.content.Context;
import android.database.Cursor;
import android.database.sqlite.SQLiteDatabase;
import java.util.ArrayList;
import java.util.List;
public class CostDAO {
    private DBHelper dbHelper;
    private SQLiteDatabase db;
    public CostDAO(Context context) {
        dbHelper = new DBHelper(context);
        db = dbHelper.getWritableDatabase();
    }
    /*添加账目*/
    public boolean addCost(CostBean cost) {
        boolean flag = false;
        if (cost != null) {
            String sql = "INSERT INTO cost(cost_type, cost_money, cost_title, cost_explain, cost_time,cost_pic,user_id) VALUES(?,?,?,?,?,?,?)";
            db.execSQL(sql, new Object[]{cost.getType(),cost.getMoney(),cost.getTitle(),cost.getExplain(), cost.getTime(),cost.getPicPath(),cost.getUserID()});
            flag = true;
        }
        return flag;
    }
    /*删除账目*/
    public void deleteCost(int costID) {
        String sql = " DELETE FROM cost WHERE cost_id = ?";
        db.execSQL(sql, new String[]{String.valueOf(costID)});
    }
    /*返回List型的查询结果*/
    public List<CostBean> getCostList(int userID,String type) {
        List<CostBean> costList = new ArrayList<>();
        String sql;
        Cursor cursor;
        //为简化逻辑,直接通过消费类型名称字符串进行判断
        if(type.equals("全部")){
```

```
            sql = "SELECT * FROM cost WHERE user_id=? ORDER BY cost_time DESC"; //按照cost_id倒序排列
            cursor = db.rawQuery(sql, new String[]{String.valueOf(userID)});
        }else{
            sql = "SELECT * FROM cost WHERE user_id=? AND cost_type=? ORDER BY cost_time DESC";
            cursor = db.rawQuery(sql, new String[]{String.valueOf(userID),type});
        }
        while (cursor.moveToNext()) {
            CostBean costBean = new CostBean();
            costBean.setCostID(cursor.getInt(cursor.getColumnIndex("cost_id")));
            costBean.setType(cursor.getString(cursor.getColumnIndex("cost_type")));
            costBean.setMoney(cursor.getFloat(cursor.getColumnIndex("cost_money")));
            costBean.setTitle(cursor.getString(cursor.getColumnIndex("cost_title")));
            costBean.setExplain(cursor.getString(cursor.getColumnIndex("cost_explain")));
            costBean.setTime(cursor.getString(cursor.getColumnIndex("cost_time")));
            costBean.setPicPath(cursor.getString(cursor.getColumnIndex("cost_pic")));
            costBean.setUserID(cursor.getInt(cursor.getColumnIndex("user_id")));
            costList.add(costBean);
        }
        cursor.close();
        return costList;
    }
    /*关闭连接*/
    public void close() {
        if (db != null) {
            db.close();
        }
    }
}
```

(3) 账目信息添加。在MainRecordFragment中对提交按钮进行事件监听,当用户点击提交按钮时,读取界面上用户输入的各个数据项并且获取到SharedPreferences中用户的id属性值,将这些数据封装到一个UserBean对象中,然后调用CostDAO中的addCost方法添加数据。在MainRecordFragment的onCreateView方法中添加如下代码:

```
sp = getContext().getSharedPreferences("user", getContext().MODE_PRIVATE);
btnAdd.setOnClickListener(new View.OnClickListener() {
    @Override
```

```java
public void onClick(View v) {
    if (TextUtils.isEmpty(edtMoney.getText().toString().trim())) {
        Toast.makeText(getActivity(), "请添加金额", Toast.LENGTH_SHORT).show();
    } else {
        CostDAO costDAO = new CostDAO(getActivity());
        CostBean bean = new CostBean(); //创建 bean 并赋值
        String time = dpTime.getYear() + "-" + String.valueOf(dpTime.getMonth() + 1) + "-" + dpTime.getDayOfMonth();
        bean.setTime(time);
        bean.setTitle(edtTitle.getText().toString());
        bean.setExplain(edtExplain.getText().toString());
        bean.setType(spType.getSelectedItem().toString());
        bean.setMoney(Float.valueOf(edtMoney.getText().toString()));
        bean.setUserID(sp.getInt("userID", 0));
        costDAO.addCost(bean);//调用方法添加赋值后的 bean
        costDAO.close();//关闭数据库连接
        Intent intent = new Intent(getActivity(), TranslucentActivity.class);
        startActivity(intent);
    }
}
});
```

　　账目添加功能已经实现。但是，由于账目显示的 Fragment 此时处于隐藏状态，当用户点击底部导航按钮"账目列表"时，仅仅是把它从隐藏状态切换到显示状态，并没有重新去读取数据表中的数据，这导致刚刚添加的账目信息虽然已经被添加到数据库中，但是此时无法显示在账目列表中。同时，添加数据后，账目添加 MainRecordFragment 中用户刚才输入的数据依然保留在界面上，没有进行清空，用户如果再次点击提交"确认"按钮，将会向数据表中再添加一组同样的账目信息，这显然会影响用户的体验。所以，应该对以上账目添加的功能进行改进，避免这两种问题出现。

**改进思路：**

　　创建一个透明的 Activity 作为数据添加成功的提示窗口。这个透明的 Activity 覆盖整个屏幕，其上放置一个"添加成功"提示按钮。它主要有两个作用：

　　（1）从 MainActivity 跳转到这个透明 Activity 时，MainActivity 将进入暂停状态。如上代码，已经添加了跳转功能在数据添加之后：

```java
Intent intent = new Intent(getActivity(), TranslucentActivity.class);
startActivity(intent);
```

　　（2）当用户点击透明 Activity 上的确认按钮时，关闭这个透明 Activity，并通过一些方式重新读取数据并更新界面，例如在 onResume 方法中重新初始化各个 Fragment、基于服务的

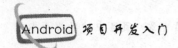

方式修改数据等。本例为简化业务逻辑，采取一个简单的策略，即直接通过 Intent 跳转到新的 MainActivity，这样每次数据更改时会创建一个新的 MainActivity 对象，程序性能受到影响。

TranslucentActivity 的实现如下：

在项目列表中新建一个"Empty Activity"，命名为 TranslucentActivity。在布局文件 activity_translucent.xml 中添加一个按钮，主要属性如下：

```xml
<Button
    android:id="@+id/btn_OK_trans"
    style="@style/BtnTheme"
    android:text="添加成功"/>
```

在 TranslucentActivity 的 Java 代码中实现事件监听，跳转并关闭当前 Activity。

```java
btnOK=findViewById(R.id.btn_OK_trans);
btnOK.setOnClickListener(new View.OnClickListener() {
    @Override
    public void onClick(View v) {
        Intent intent = new Intent(TranslucentActivity.this, MainActivity.class);
        startActivity(intent);
        TranslucentActivity.this.finish();
    }
});
```

为 TranslucentActivity 设置透明样式。

在 styles.xml 中添加一个自定义样式：

```xml
<style name="TranslucentTheme" parent="Theme.AppCompat.Light.NoActionBar">
    <item name="android:windowNoTitle">true</item>
    <item name="android:windowBackground">@color/transparent</item>
    <item name="android:windowIsTranslucent">true</item>
</style>
```

其中，颜色"@color/transparent"在 colors.xml 中添加：

```xml
<color name="transparent">#90000000</color>
```

在这个样式文件中，"android:windowIsTranslucent"使得当前 Activity 打开后处于透明状态，同时，它的前一个 Activity 会停留在暂停状态，不会被停止或者销毁。

在配置文件 AndroidManifest.xml 中为 TranslucentActivity 添加主题：

```xml
<activity android:name=".TranslucentActivity"
```

android:theme="@style/TranslucentTheme"/>

运行程序,打开"记一笔"功能,录入信息,点击"确认"按钮,可以看到运行效果如图5.4所示。

图5.4 添加账目

### 4. 账目信息列表显示

**[项目案例16] 在AccountBook项目中实现账目信息显示功能**

在上一章对项目界面进行实现时,已经在AccountAllFragment中通过自定义的initData()方法模拟了一组账目信息的数据对象,添加到List<CostBean>中作为数据源提供给了列表进行显示。在有了数据操作类costDAO后,就可以调用它的getCostList()方法直接得到一个List<CostBean>列表对象作为数据源。

getCostList()方法有两个参数,第一个参数是用户id,用来标识是哪一个用户,第二个参数是查询的类别,它要与数据库查询语句对应,可选参数为"全部""用餐""服装""交通""娱乐"。

例如:在AccountAllFragment中,需要获取所有类型的数据,其实现代码为:

```
private void initData() {
    costDAO = new CostDAO(getActivity());
    sp = getContext().getSharedPreferences("user", getContext().MODE_PRIVATE);
    int userID=sp.getInt("userID", 0);
    list = costDAO.getCostList(userID,"全部");
    costDAO.close();
}
```

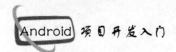

在 AccountEatFragment 等其他四个 Fragment 中，只需要获取对应类型的数据，其实现代码分别为：

list = costDAO.getCostList(userID,"用餐");

list = costDAO.getCostList(userID,"服装");

list = costDAO.getCostList(userID,"交通");

list = costDAO.getCostList(userID,"娱乐");

运行程序，添加数据后，账目列表显示如图 5.5 所示。

图 5.5　账目列表显示

### 5. 账目信息删除

**[项目案例 17]　在 AccountBook 项目中实现账目删除功能**

用户长按一条账目的列表项，弹出一个提示框，用户点击"删除"，则该条目从列表中删除，同时从数据表中删除，用户点击"取消"，则不删除数据。

由于在 RecyclerView 实现时，无法直接获取当前用户选择的某一个列表项位置，所以为了实现 RecyclerView 列表项的删除功能，可以在适配器 AccountItemAdapter 类的 onBindViewHolder() 方法中添加长按列表项的上下文菜单事件响应，当用户在菜单上点击时，再响应 onMenuItemClick 事件，首先删除 List 列表中的数据，然后通过 notifyItemRemoved() 方法对列表显示进行刷新。同时，调用数据操作类 CostDAO 中的 deleteCost() 方法通过传入 id 值删除该条数据。注意，此时的 position 和 bean 变量需要声明成 final 类型。在 AccountItemAdapter 类中对 onBindViewHolder() 方法的修改和代码添加如下：

```
public void onBindViewHolder(ViewHolder viewHolder, int i) {
final int position =i;
final CostBean bean = list.get(i);
viewHolder. itemView. setOnCreateContextMenuListener(new  View.OnCreateContextMenuLis-
```

```
tener(){
    @Override
    public void onCreateContextMenu(ContextMenu menu, View v, ContextMenu.ContextMenuInfo menuInfo) {
        MenuItem delete = menu.add(0, 0, 0, "删除");
        delete.setOnMenuItemClickListener(new MenuItem.OnMenuItemClickListener() {
            @Override
            public boolean onMenuItemClick(MenuItem item) {
                list.remove(list.get(position));
                notifyItemRemoved(position);
                costDAO = new CostDAO(context);
                costDAO.deleteCost(bean.getCostID());
                Intent intent = new Intent(context, TranslucentActivity.class);
                context.startActivity(intent);
                return false;
            }
        });
    }
});
}
```

运行程序,长按列表项,可以看到弹出删除菜单,点击该菜单项,一条数据就会被删除。程序运行效果如图5.6所示。请注意在本段代码中,删除数据后还加入了以下两句:

```
Intent intent = new Intent(context, TranslucentActivity.class);
context.startActivity(intent);
```

这是因为由于ViewPager的缓存机制会默认缓存当前显示视图左右两边各一个视图,即当前视图显示时,它左右两边最近的视图也会被创建出来,以提高滑屏加载效率。在本项目中,当显示"全部"列表时,"用餐"列表也已经创建好,这导致用户如果删除"全部"列表中的一条用餐数据,虽然数据库中也进行了删除,但是"用餐"界面没有进行数据更新操作,因此用户点击"用餐"导航标签时,刚才删除的数据还在用餐列表中。为了避免这种情况出现,本项目依然采用数据添加操作时的实现策略:删除数据后,跳转到一个透明Activity进行操作确认,然后关闭透明Activity,跳转到一个新的MainActivity。

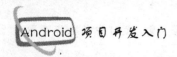

图 5.6 删除账目

### 5.2.3 图表分析功能实现

**[项目案例18] 在 AccountBook 项目中实现一个图表统计功能**

本项目以饼状图显示各类消费占比为例。首先从数据表中查询用户分类消费的单项总额，如全部用餐费用、全部服装费用等，然后计算出各类单项消费总额在全部消费中的百分比，将各个百分比值作为 PieChart 的显示数据。

（1）在数据操作 CostDAO 类中添加一个查询单项总额的方法。该方法通过用户 id 和分项名称作为参数，利用 SQL 中的 SUM 函数直接返回一列数据的求和结果。

```
/*查询指定用户分项消费总额,type为分项名称,如"用餐"*/
public float getTotal(int userID,String type){
    float total;
    String sql = "SELECT SUM(cost_money) FROM cost WHERE user_id=? AND cost_type=?";//SUM为累加函数
    Cursor cursor = db.rawQuery(sql, new String[]{String.valueOf(userID),type});
    cursor.moveToFirst();
    total = cursor.getFloat(0);
    return total;
}
```

（2）在 MainChartFragment 中调用 getTotal 方法，获得各个分类的消费总额，然后计算百分比，并赋值给 PieChart。修改 MainChartFragment 代码如下：

```
package cc.turbosnail.accountbook;
import android.content.SharedPreferences;
import android.graphics.Color;
import android.os.Bundle;
```

```java
import android.support.v4.app.Fragment;
import android.view.LayoutInflater;
import android.view.View;
import android.view.ViewGroup;
import com.github.mikephil.charting.charts.PieChart;
import com.github.mikephil.charting.components.Description;
import com.github.mikephil.charting.data.PieData;
import com.github.mikephil.charting.data.PieDataSet;
import com.github.mikephil.charting.data.PieEntry;
import com.github.mikephil.charting.formatter.PercentFormatter;
import java.text.DecimalFormat;
import java.util.ArrayList;
public class MainChartFragment extends Fragment {
    private PieChart chartAccount;
    private CostDAO costDAO;
    private float eat,clothes,traffic,play,total;
    public View onCreateView(LayoutInflater inflater, ViewGroup container,
                Bundle savedInstanceState) {
        View view = inflater.inflate(R.layout.main_chart, container, false);
        initData();
        chartAccount = view.findViewById(R.id.chart_account);
        chartAccount.setDrawHoleEnabled(false);//设置为实心
        Description description = new Description();
        description.setText("消费分类统计图");
        chartAccount.setDescription(description);//设置右下方文本标签内容
        DecimalFormat df=new DecimalFormat("0.00");
        float per1= Float.parseFloat(df.format(eat/total*100));
        float per2= Float.parseFloat(df.format(clothes/total*100));
        float per3= Float.parseFloat(df.format(traffic/total*100));
        float per4= Float.parseFloat(df.format(play/total*100));
        ArrayList<PieEntry> entries = new ArrayList<>();//数据列表
        entries.add(new PieEntry(per1,"用餐"));//(数据,分类名)
        entries.add(new PieEntry(per2,"服装"));
        entries.add(new PieEntry(per3,"交通"));
        entries.add(new PieEntry(per4,"娱乐"));
        PieDataSet dataSet = new PieDataSet(entries, "");//构建PieDataSet
        dataSet.setValueTextSize(18f);//设置字体大小
        ArrayList<Integer> colors = new ArrayList<>();//定义颜色列表
        colors.add(Color.rgb(205, 205, 205));
```

```java
            colors.add(Color.rgb(114, 188, 223));
            colors.add(Color.rgb(255, 123, 124));
            colors.add(Color.rgb(57, 135, 200));
            dataSet.setColors(colors);//为 PieDataSet 设置颜色
            PieData pieData = new PieData(dataSet);//构建 PieData
            pieData.setValueFormatter(new PercentFormatter());//百分比显示,默认直接显示数值
            chartAccount.setData(pieData);//为图表添加 PieData 数据
            chartAccount.invalidate();//刷新
            return view;
        }
        private void initData() {
            costDAO = new CostDAO(getActivity());
        SharedPreferences sp = getContext(). getSharedPreferences("user", getContext(). MODE_PRIVATE);
            int userID=sp.getInt("userID",0);
            eat= costDAO.getTotal(userID,"用餐");
            clothes= costDAO.getTotal(userID,"服装");
            traffic= costDAO.getTotal(userID,"交通");
            play= costDAO.getTotal(userID,"娱乐");
            total=eat+clothes+traffic+play;
            costDAO.close();
        }
    }
```

运行程序,可以看到,打开"图表分析"界面时,饼状图中显示的是数据表中存储的各类消费总额所占的百分比。通过获取不同的数据(例如每月的消费数据)和不同的图形(例如折线图、柱状图),可以进一步地实现丰富的数据统计分析功能。

## 5.3 File文件操作

Android文件操作和标准的Java文件操作基本一致,支持File类和各种数据流操作,但是在文件存储路径的管理上,Android与标准的Java不同,它提供了三种存储路径,不同存储路径的获取方法也不同。

(1)内置存储。移动设备生产时内置的存储空间。应用程序安装后,内置存储会为它分配一个固定的文件存储目录,这个目录默认是该程序私有的,没有root权限的用户也无法通过文件管理器直接查看,程序卸载时这个目录会被删除,通过设备的"清除数据""清除缓存"也可以删除其中的数据。程序敏感数据可以存储在这个目录下。

public File getFilesDir()，获取内置存储下的文件根目录。

public abstract File getCacheDir()，获取内置存储下的缓存根目录,内置存储空间不足时会被自动清除。

Context实例还提供如下方法来进行内置存储文件操作：

- public FileInputStream openFileInput(String name)，打开名为name的文件。返回FileInputStream对象,可以基于这个对象进行文件的读入操作,如读取文件中的内容。

- public FileOutputStream openFileOutput(String name, int mode)，打开名为name的文件,如果该文件不存在,则新建这个文件。mode常量包括MODE_APPEND(写入的内容将跟在被追加的文件末尾而不是覆盖原内容)、MODE_PRIVATE(文件私有)。返回FileOutputStream对象,可以基于这个对象进行文件的写出操作,如保存、追加等。

- public String[] fileList()，以字符串数组的形式返回当前目录下的所有文件名。

- public boolean deleteFile(String name)，删除一个名为name的文件,删除成功返回true,失败返回false。

（2）外置扩展存储。外置存储设备包括SD卡、TF卡等。外置扩展存储是Android在外置存储设备上给每个应用程序分配的一个固定存储目录。由于是在外置存储设备上,因此这个目录中的文件可以被其他程序访问,用户也可直接查看。和内置存储目录一样,程序卸载时这个目录会被删除,通过设备的"清除数据""清除缓存"也可以删除其中的数据。程序的非敏感数据建议都存放在这个目录下。Contex提供获取外置扩展存储目录的方法如下：

public File getExternalFilesDir()，获取外置扩展存储下的根目录。

public File getExternalCacheDir()，获取外置扩展存储下的缓存目录,内置存储空间不足时会被自动清除。

因为外置存储不一定存在,所以在使用时需要先通过以下语句判断是否存在外存储设备：

if (Environment.getExternalStorageState().equals(Environment.MEDIA_MOUNTED))

（3）外置公共存储。外置公共存储是除了外置扩展存储之外的外置存储部分,用户可以从根目录开始自定义存储目录,这里的数据对其他应用程序和用户都是公开的,在程序卸载时不会自动卸载。开发者需要在配置文件中申请授权才能使用外置公共存储,获取SD卡文件操作权限需要在AndroidManifest.xml配置文件的根目录<manifest>标签中添加如下代码：

```
<uses-permission android:name="android.permission.READ_EXTERNAL_STORAGE"/>
<uses-permission android:name="android.permission.WRITE_EXTERNAL_STORAGE"/>
```

在Android6.0以后,Google增强了文件操作的安全性,要求必须经过用户授权才能访问外置公共存储,所以除了需要在AndroidManifest.xml中进行声明之外,还需要在程序中进行动态文件授权,可以定义如下方法进行授权：

```
public static void verifyStoragePermissions(Activity activity) {
    //读写权限
    String[] PERMISSIONS_STORAGE = {Manifest.permission.READ_EXTERNAL_STORAGE,
        Manifest.permission.WRITE_EXTERNAL_STORAGE};
    try {
        //检测是否有写的权限
        for (int i = 0; i < PERMISSIONS_STORAGE.length; i++) {
            int permission = ActivityCompat.checkSelfPermission(activity, PERMISSIONS_STORAGE[i]);
            int REQUEST_PERMISSION_CODE = i;//返回码,onRequestPermissionsResult 方法中判断时使用
            if (permission != PackageManager.PERMISSION_GRANTED) {
                //申请权限,会自动弹出对话框
                ActivityCompat. requestPermissions(activity, new String[] {PERMISSIONS_STORAGE[i]}, REQUEST_PERMISSION_CODE);
            }
        }
    } catch (Exception e) {
        e.printStackTrace();
    }
}
```

在 Activity 的 onCreate 方法中,调用该方法进行授权:

verifyStoragePermissions(this);

此时会弹出一个对话框由用户选择是否运行程序访问外置公共存储。示意图如图5.7所示。

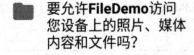

图5.7 外置公共存储动态授权

获取外置存储根目录的方法由 Environment 提供,其方法如下:

public static File getExternalStoragePublicDirectory（String type），参数 type 是文件目录名，可以是 DIRECTORY_MUSIC、DIRECTORY_PODCASTS、DIRECTORY_RINGTONES、DIRECTORY_ALARMS、DIRECTORY_NOTIFICATIONS、DIRECTORY_PICTURES、DIRECTORY_MOVIES、DIRECTORY_DOWNLOADS、DIRECTORY_DCIM 等常量之一，也可以用户自定义，但是对于不存在的文件，一定要通过 mkdirs( )方法创建目录。示例如下：

```
if (Environment.getExternalStorageState().equals(Environment.MEDIA_MOUNTED)) {
    File path = Environment.getExternalStoragePublicDirectory("myFolder");
        if(!path.exists()){
            path.mkdirs();
        }
    File file = new File(path, "data.txt");
    //……//文件操作
}
```

【例5.1】文件操作，访问内置存储、外置扩展存储、外置公共存储示例。

（1）在 Android Studio 中新建一个项目，项目名为 FileDemo。

（2）打开项目目录中的 AndroidManifest.xml 配置文件，在根目录<manifest>标签中添加如下授权代码：

```
<uses-permission android:name="android.permission.READ_EXTERNAL_STORAGE"/>
<uses-permission android:name="android.permission.WRITE_EXTERNAL_STORAGE"/>
```

（3）在 activity_main.xml 布局文件中添加三组用来显示存储路径、文件内容等的文本标签，并添加三个按钮分别用来在事件响应中进行三种文件的读写操作。界面布局如图 5.8 所示。

图 5.8 文件操作示例界面设计

（4）在 MainActivity 中封装各类文件读、写的方法，以及外置公共存储动态授权的方法，然后在 onCreate 方法中进行方法调用，并在界面上输出测试结果。

```java
package cc.turbosnail.filedemo;
import android.Manifest;
import android.app.Activity;
import android.content.pm.PackageManager;
import android.os.Bundle;
import android.os.Environment;
import android.support.v4.app.ActivityCompat;
import android.support.v7.app.AppCompatActivity;
import android.view.View;
import android.widget.Button;
import android.widget.TextView;
import java.io.BufferedReader;
import java.io.File;
import java.io.FileInputStream;
import java.io.FileNotFoundException;
import java.io.FileOutputStream;
import java.io.IOException;
import java.io.InputStreamReader;
public class MainActivity extends AppCompatActivity {
    private TextView tvInsideContent, tvInsidePath, tvOutsideContent,
            tvOutsidePath, tvExternalContent, tvExternalPath;
    private Button btnInside, btnOutside, btnExternal;
    @Override
    protected void onCreate(Bundle savedInstanceState) {
        super.onCreate(savedInstanceState);
        setContentView(R.layout.activity_main);
        verifyStoragePermissions(this);
        initView();
        btnInside.setOnClickListener(new View.OnClickListener() {
            @Override
            public void onClick(View view) {
                //内置存储
                writeInsideFile("All grown-ups were once children-- although few of them remember it.");
                tvInsideContent.setText("存储内容:" + readInsideFile());
                tvInsidePath.setText("存储路径:" + MainActivity.this.getFilesDir().getAbsolutePath
```

```java
            ());//内置根目录
        }
    });
    btnOutside.setOnClickListener(new View.OnClickListener() {
        @Override
        public void onClick(View view) {
            //外置扩展存储
            writeExternalFile("It is only with the heart that one can see rightly; what is essential is invisible to the eye.");
            tvOutsideContent.setText("存储内容:" + readExternalFile());
            tvOutsidePath.setText("存储路径:" + MainActivity.this.getExternalFilesDir("myFolder2").getAbsolutePath());//SD卡根目录
        }
    });
    btnExternal.setOnClickListener(new View.OnClickListener() {
        @Override
        public void onClick(View view) {
            //外置公共存储
            writeExternalPubFile("What makes the desert beautiful is that somewhere it hides a well...");
            tvExternalContent.setText("存储内容:" + readExternalPubFile());
            tvExternalPath.setText("存储路径:"+Environment.getExternalStoragePublicDirectory("myFolder3").getAbsolutePath());//SD卡根目录下自定义目录
        }
    });
}
public static void verifyStoragePermissions(Activity activity) {
    //读写权限
    String[] PERMISSIONS_STORAGE = {Manifest.permission.READ_EXTERNAL_STORAGE,
            Manifest.permission.WRITE_EXTERNAL_STORAGE};
    try {
        //检测是否有写的权限
        for (int i = 0; i < PERMISSIONS_STORAGE.length; i++) {
            int permission = ActivityCompat.checkSelfPermission(activity, PERMISSIONS_STORAGE[i]);
            int REQUEST_PERMISSION_CODE = i;//返回码,onRequestPermissionsResult方法中
```

判断时使用

```java
            if (permission != PackageManager.PERMISSION_GRANTED) {
                //申请权限,会自动弹出对话框
                ActivityCompat.requestPermissions(activity, new String[]{PERMISSIONS_STORAGE[i]}, REQUEST_PERMISSION_CODE);
            }
        }
    } catch (Exception e) {
        e.printStackTrace();
    }
}
/*内部存储:写出*/
public void writeInsideFile(String content) {
    try {
        FileOutputStream fos = openFileOutput("data1.txt", MODE_PRIVATE);
        fos.write(content.getBytes());
        fos.close();
    } catch (IOException e) {
        e.printStackTrace();
    }
}
/*内部存储:读入*/
public String readInsideFile() {
    String data = "";
    FileInputStream fis;
    try {
        fis = openFileInput("data1.txt");
        int len = fis.available();
        byte[] buffer = new byte[len];
        fis.read(buffer);
        data = new String(buffer);
        fis.close();
    } catch (FileNotFoundException e) {
        e.printStackTrace();
    } catch (IOException e) {
        e.printStackTrace();
```

        }
        return data;
    }
    /*外置扩展存储:写出*/
    public void writeExternalFile(String content) {
        if (Environment.getExternalStorageState().equals(Environment.MEDIA_MOUNTED)) {//判断外部设备是否可用
            File path = this.getExternalFilesDir("myFolder2");
            File file = new File(path, "data2.txt");
            FileOutputStream fos;
            try {
                fos = new FileOutputStream(file);
                fos.write(content.getBytes());
                fos.close();
            } catch (FileNotFoundException e) {
                e.printStackTrace();
            } catch (IOException e) {
                e.printStackTrace();
            }
        }
    }
    /*外置扩展存储:读入*/
    public String readExternalFile() {
        String data = "";
        if (Environment.getExternalStorageState().equals(Environment.MEDIA_MOUNTED)) {
            File path = this.getExternalFilesDir("myFolder2");
            File file = new File(path, "data2.txt");
            FileInputStream fis;
            try {
                fis = new FileInputStream(file);
                BufferedReader buffer = new BufferedReader(new InputStreamReader(fis));
                String temp;
                while ((temp = buffer.readLine()) != null) {
                    data += temp;
                }
                buffer.close();

```java
                fis.close();
            } catch (FileNotFoundException e) {
                e.printStackTrace();
            } catch (IOException e) {
                e.printStackTrace();
            }
        }
        return data;
    }
    /*外置公共存储:写出*/
    public void writeExternalPubFile(String content) {
        if (Environment.getExternalStorageState().equals(Environment.MEDIA_MOUNTED)) {//判断外部设备是否可用
            File path = Environment.getExternalStoragePublicDirectory("myFolder3");
            if (!path.exists()) {
                path.mkdirs();
            }
            File file = new File(path, "data3.txt");
            FileOutputStream fos;
            try {
                fos = new FileOutputStream(file);
                fos.write(content.getBytes());
                fos.close();
            } catch (FileNotFoundException e) {
                e.printStackTrace();
            } catch (IOException e) {
                e.printStackTrace();
            }
        }
    }
    /*外置公共存储:读入*/
    public String readExternalPubFile() {
        String data = "";
        if (Environment.getExternalStorageState().equals(Environment.MEDIA_MOUNTED)) {
            File path = Environment.getExternalStoragePublicDirectory("myFolder3");
            File file = new File(path, "data3.txt");
```

```
        FileInputStream fis;
        try {
            fis = new FileInputStream(file);
            BufferedReader buffer = new BufferedReader(new InputStreamReader(fis));
            String temp;
            while ((temp = buffer.readLine()) != null) {
                data += temp;
            }
            buffer.close();
            fis.close();
        } catch (FileNotFoundException e) {
            e.printStackTrace();
        } catch (IOException e) {
            e.printStackTrace();
        }
    }
    return data;
}
/*初始化界面组件*/
private void initView() {
    tvInsideContent = findViewById(R.id.tv_inside_content);
    tvInsidePath = findViewById(R.id.tv_inside_path);
    tvOutsideContent = findViewById(R.id.tv_outside_content);
    tvOutsidePath = findViewById(R.id.tv_outside_path);
    tvExternalContent = findViewById(R.id.tv_external_content);
    tvExternalPath = findViewById(R.id.tv_external_path);
    btnInside = findViewById(R.id.btn_inside);
    btnOutside = findViewById(R.id.btn_outside);
    btnExternal = findViewById(R.id.btn_external);
}
}
```

首次运行程序时,会弹出动态授权对话框,选择"允许"可进入程序界面,分别点击三个按钮,可以看到运行效果如图5.9所示。

图 5.9　文件操作示例运行效果

　　本例的重点是封装了三种文件读、写操作的相关方法,在涉及文件操作的应用开发中,读者可以根据数据存储的需要,选择具体的方法进行调用和扩展。

# 第6章 使用手机相册——ContentProvider

在个人记账本项目中,如果希望从手机上已有的图片中读取一张图片作为账目列表的图标,即使用手机存储卡上的图片文件,则可以按照文件操作的方式先遍历整个内置存储和外置存储的各层文件目录,然后过滤出相应格式的图片文件,再进行相关读写操作,这样的业务实现既费时又繁杂,如果此时可以由记账本程序直接获取系统图库程序的文件列表,就可以很方便地得到所需要的图片文件,为此,这就需要在一个应用程序中访问另一个应用程序的数据。当需要在一个应用程序中访问别的应用程序中的数据时,Android提供了ContentProvider来实现,应用程序通过ContentProvider将部分数据进行公开,这样其他应用程序就可以直接访问这些公开的数据。Android系统本身提供了大量的公开数据,如图片、音视频、通信录等,这些数据来自于媒体库、手机通信录等程序。同时,Android也允许开发者自定义公开数据以便其他程序访问。

ContentProvider是Android的四大基础组件之一,它为不同应用程序之间的数据共享提供服务,ContentProvider类提供数据,ContentResolver类负责访问数据,它们作为一组"中间人",让两个独立的应用程序实现数据共享,无论应用程序底层数据存储采用何种方式,ContentProvide使得外界对数据的访问形式都是统一的。

## 6.1 系统提供的ContentProvider

Android系统程序提供了大量的ContentProvider,例如手机通信录里的联系人信息、图片库信息等,开发者可以在自己的程序中通过ContentResolver直接访问,共享这些数据。ContentResolver的用法如下:

(1)获取Contex的ContentResolver对象。

ContentResolver contentResolver = getContentResovler()

(2)调用ContentResolver的方法进行数据操作,ContentResolver提供了一系列数据操作的方法,这些方法一般通过Uri(Uniform Resource Identifier,统一资源标识符)访问对应ContentProvider提供的数据,常用的方法如下:

- ☞ public final Uri insert (Uri url, ContentValues values)。values是待插入数据,ContentValues类型。

- public final int delete（Uri url, String where, String[] selectionArgs）。where是条件字符串，如"id = 1"。selectionArgs为其他条件组成的数组。
- public final Cursor query（Uri uri, String[] projection, String selection, String[] selectionArgs, String sortOrder）。projection是要返回的列构成的数组，例如：projection=new String[]{android.provider.ContactsContract.Contacts.DISPLAY_NAME}要返回的是NAME列。Selection是查询条件（where）字符串。selectionArgs是其他查询条件的字符串数组。sortOrder是排序规则字符串。
- public final int update（Uri uri, ContentValues values, String where, String[] selectionArgs）。values是待更新的数据，ContentValues是类型。where是条件语句字符串。selectionArgs是其他条件字符串组成的数组。

常见的系统ContentProvider包括多媒体文件的Media Provider、手机通信录的Contacts Provider、日历的Calendar Provider等。

### 6.1.1 访问手机通信录信息

通信录（Contacts）是Android系统自带的应用程序，它通过ContentProvider的形式允许开发者访问通信录数据。访问通信录需首先在AndroidManifest.xml文件中进行授权：

```
<!-- 读联系人权限 -->
<uses-permission android:name="android.permission.READ_CONTACTS"/>
<!-- 写联系人权限 -->
<uses-permission android:name="android.permission.WRITE_CONTACTS"/>
```

在Android 6.0之后还需要进行动态授权，授权代码的写法见本节示例代码。

访问通信录主要涉及的三张数据表为：

- raw_contacts表：联系人数据表，一个联系人占一行，允许同名的联系人。
- contacts表：联系人表，存储联系人的姓名、头像、最后通话时间等信息，同名联系人会被合并，通信录软件中实际显示的联系人。
- data表：联系人信息数据。通过raw_contact_id外键与raw_contacts关联。

在进行查询操作时，可以直接通过Uri获取对应结果或者结果集，如果是结果集，则通过遍历的形式获取数据，当需要查看所有列的名称时，可以通过Cursor类的getColumnnames()方法返回一个字符串类型的列名数组，以方便查找数据，例如：

String[] columnArray = cursor.getColumnnames()

在进行联系人信息修改、删除、添加时，可以根据Uri访问相关表，获取字段名称，然后构建ContentValues参数，再调用ContentResolver中对应的方法完成操作。一些常用的Uri如下：

- ContactsContract.Contacts.CONTENT_URI，联系人Uri，对应content://com.android.contacts/contacts。
- ContactsContract.CommonDataKinds.Phone.CONTENT_URI，联系人电话Uri，对应content://com.android.contacts/data/phones。

- ContactsContract.CommonDataKinds.Email.CONTENT_URI，联系人 Email 的 Uri，content://com.android.contacts/data/emails。
- ContactsContract.CommonDataKinds.StructuredPostall.CONTENT_URI，联系人地址 Uri，对应 content://com.android.contacts/data/postals。
- ContactsContract.Data.CONTENT_URI，data 表，对应 content://com.android.contacts/data。
- content://contacts/people，所有联系人。
- content://contacts/people/x，某个联系人，x 为 id。

**【例6.1】** 在一个 Android 项目中读取手机通信录中所有联系人的信息，并输出到界面上显示出来。

（1）在 Android Studio 中新建一个项目，项目名为 ContactsDemo。

（2）打开项目目录中的 AndroidManifest.xml 配置文件，在根目录<manifest>标签中添加如下授权代码：

```xml
<!-- 读联系人权限 -->
<uses-permission android:name="android.permission.READ_CONTACTS"/>
<!-- 写联系人权限 -->
<uses-permission android:name="android.permission.WRITE_CONTACTS"/>
```

（3）在 activity_main.xml 布局文件中放置一个文本标签，id 为 tv_contacts，用来显示读取出来的数据。

（4）在 MianActivity 中实现一个 contactsInfo() 方法来读取通信录中的数据，并输出显示。然后根据动态授权情况，调用 contactsInfo() 方法。代码清单如下：

```java
package cc.turbosnail.contactsdemo;
import android.Manifest;
import android.content.ContentResolver;
import android.content.pm.PackageManager;
import android.database.Cursor;
import android.net.Uri;
import android.os.Build;
import android.provider.ContactsContract;
import android.support.v4.app.ActivityCompat;
import android.support.v4.content.ContextCompat;
import android.support.v7.app.AppCompatActivity;
import android.os.Bundle;
import android.util.Log;
import android.widget.TextView;
import android.widget.Toast;
```

```java
public class MainActivity extends AppCompatActivity {
    private TextView tvContacts;//用来显示读取出来的数据
    @Override
    protected void onCreate(Bundle savedInstanceState) {
        super.onCreate(savedInstanceState);
        setContentView(R.layout.activity_main);
        tvContacts = findViewById(R.id.tv_contacts);
        /*动态授权*/
        if (Build.VERSION.SDK_INT >= 23) {
            int request = ContextCompat. checkSelfPermission(this, Manifest. permission. READ_CONTACTS);//检查通信录读取权限
            if (request != PackageManager.PERMISSION_GRANTED) {
                ActivityCompat. requestPermissions(this, new String[] {Manifest. permission. READ_CONTACTS}, 1);//授权,1 为自定义的返回码
            } else {
                contactsInfo();//已经授权,直接读取通信录数据
            }
        }
    }
    /*授权的回调方法,可根据返回值,进行相应处理*/
    @Override
    public void onRequestPermissionsResult(int requestCode, String[] permissions, int[] grantResults) {
        super.onRequestPermissionsResult(requestCode, permissions, grantResults);
        //通过返回码进行判断
        if (requestCode == 1) {
            if (grantResults[0] == PackageManager.PERMISSION_GRANTED) {
                Toast.makeText(this, "权限申请成功", Toast.LENGTH_SHORT).show();
                contactsInfo(); //读取通信录信息
            } else if (grantResults[0] == PackageManager.PERMISSION_DENIED) {
                Toast.makeText(this, "权限申请失败,用户拒绝权限", Toast.LENGTH_SHORT).show();
            }
        }
    }
    private void contactsInfo() {
        Uri uri = ContactsContract.Contacts.CONTENT_URI;
        ContentResolver contentResolver = getContentResolver();
        Cursor cursor = contentResolver.query(uri, null, null, null, null);
        String message = "";
```

```
            while (cursor.moveToNext()) {
                String contactId = cursor.getString(cursor.getColumnIndex(ContactsContract.Contacts._ID));
                String name = cursor.getString(cursor.getColumnIndex(ContactsContract.Contacts.DISPLAY_NAME));
                message = message + "ID: " + contactId + ";" + "Name: " + name + ";";
                Cursor phoneCursor = contentResolver.query(ContactsContract.CommonDataKinds.Phone.CONTENT_URI, null, ContactsContract.CommonDataKinds.Phone.CONTACT_ID + " = " + contactId, null, null);
                while (phoneCursor.moveToNext()) {
                    String phoneNumber = phoneCursor.getString(phoneCursor.getColumnIndex(ContactsContract.CommonDataKinds.Phone.NUMBER));
                    message = message + "Phone: " + phoneNumber + ";";
                }
                phoneCursor.close();
                Cursor emailCursor = contentResolver.query(ContactsContract.CommonDataKinds.Email.CONTENT_URI, null, ContactsContract.CommonDataKinds.Email.CONTACT_ID + " = " + contactId, null, null);
                while (emailCursor.moveToNext()) {
                    String email = emailCursor.getString(emailCursor.getColumnIndex(ContactsContract.CommonDataKinds.Email.DATA));
                    message = message + "Email: " + email + ";";
                }
                emailCursor.close();
                message = message + "\n";
            }
            cursor.close();
            Log.i("联系人信息", message);
            tvContacts.setText("联系人信息\n" + message);
        }
}
```

在手机模拟器的通信录中添加几个联系人,如"Jim""Tom",然后运行本程序,在获得授权之后,可以看到输出联系人信息如图6.1所示。

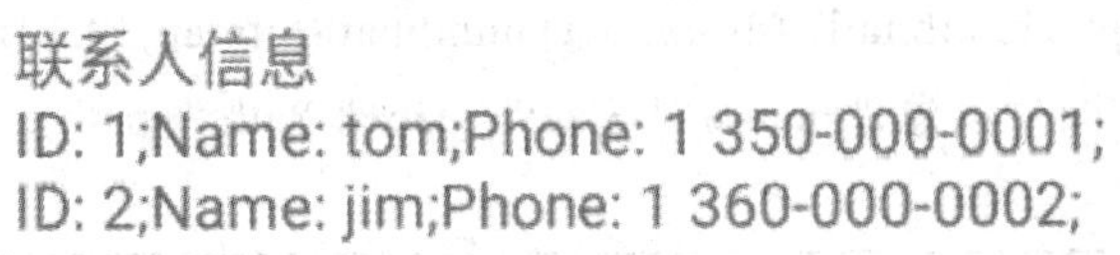

联系人信息
ID: 1;Name: tom;Phone: 1 350-000-0001;
ID: 2;Name: jim;Phone: 1 360-000-0002;

图6.1 读取手机通信录数据

### 6.1.2 访问手机多媒体信息

Android系统为多媒体库提供了ContentProvider，当需要访问手机上的多媒体信息（图片、音频、视频）时，可以通过对应的Uri由ContentResolver直接得到信息列表，包含名称、格式、路径等信息。多媒体信息常见的Uri如下：

- MediaStore.Images.Media.EXTERNAL_CONTENT_URI 外置存储中的图片Uri。
- MediaStore.Images.Media.INTERNAL_CONTENT_URI 内置存储中的图片Uri。
- MediaStore.Audio.Media.EXTERNAL_CONTENT_URI 外置存储中的音频Uri。
- MediaStore.Audio.Media.INTERNAL_CONTENT_URI 内置存储中的音频Uri。
- MediaStore.Video.Media.EXTERNAL_CONTENT_URI 外置存储中的视频Uri。
- MediaStore.Video.Media.INTERNAL_CONTENT_URI 内置存储中的视频Uri。

多媒体ContentProvider提供的只是多媒体数据的一个基本信息表，要得到实际文件，还需要根据路径读取本地文件，当需要读取的是外置存储中的内容时，需要在AndroidManifest.xml文件中进行授权：

```
<uses-permission android:name="android.permission.READ_EXTERNAL_STORAGE"/>
<uses-permission android:name="android.permission.WRITE_EXTERNAL_STORAGE"/>
```

在Android 6.0之后还需要进行动态授权，授权代码的写法见本书文件操作相关章节。一段读取内置存储中的图片文件的示例代码如下：

```
Uri uri = Uri.parse("MediaStore.Images.Media.INTERNAL_CONTENT_URI");
ContentResolver contentResolver = getContentResolver();
Cursor cursor = contentResolver.query(uri, null, null, null, null);
while(cursor.moveToNext()){
    //获取图片的名称
    String name=cursor.getString(cursor.getColumnIndex(Media.DISPLAY_NAME));
    //获取图片的大小
    String size=cursor.getString(cursor.getColumnIndex(Media.SIZE));
    //获取图片的路径
    String path=cursor.getString(cursor.getColumnIndex(Media.DATA));
    //获取原始图片
    Bitmap bitmap=BitmapFactory.decodeFile(path);
    //获取缩略图
    Bitmap smallBitmap=ThumbnailUtils.extractThumbnail(bitmap,30,30);
    Log.i("图片信息:","图片信息:"+name+" Size:"+size+" Path:"+path);
}
```

在应用开发中可以自行封装一个图片信息处理类，将遍历出来的每一组图片信息存储在图片信息类的对象中,再把这些对象放置到一个链表中,就可以根据应用程序需要使用这些图片信息数据了。

## 6.2 自定义ContentProvider

除了使用Android系统已经定义的ContentProvider之外,我们还可以把自己的应用程序数据封装成ContentProvider,允许其他应用程序通过ContentResolver来使用。自定义ContentProvider的步骤如下:

(1)创建ContentProvider的子类。

重写insert、delete、query、update等方法,这些方法实际上会由ContentResolver通过同名的方法调用执行,它还有一个onCreate方法,会在ContentResolver第一次访问它时执行一次。因此,实际数据操作的代码可以分别放在对应的方法里,初始化操作的代码可以放在onCreate方法里。其代码框架如下:

```
public class MyContentProvider extends ContentProvider {
    public boolean onCreate() {
        return false;}
     public Cursor query(Uri uri, String[] projection, String selection, String[] selectionArgs, String sortOrder) {
        return null;}
    public String getType(Uri uri) {
        return null;}
    public Uri insert(Uri uri, ContentValues values) {
        return null;}
    public int delete(Uri uri, String selection, String[] selectionArgs) {
        return 0;}
    public int update(Uri uri, ContentValues values, String selection, String[] selectionArgs) {
        return 0;}
}
```

(2)注册自定义ContentProvider并提供Uri。

为AndroidManifest.xml文件的<application>元素添加<provider>标签进行注册,<provider>有三个主要的属性:

name:自定义ContentProvider类的类名。

authorities:为自定义的ContentProvider类指定一个Uri,这个Uri就是ContentResolver访问它的"地址",一般由类的小写全称组成,以保证唯一性,如cc.turbosnail.myprovider。

exported:true/false,true表示允许当前ContentProvider被访问,false表示不允许访问。

一个ContentProvider子类在的注册如下:

```xml
<application>
<provider
    android:name="MyProvider"
    android:authorities="cc.turbosnail.myprovider"
    android:exported="true"/>
</application>
```

(3) 通过ContentResolver访问ContentProvider。

这个过程和访问Android系统的ContentProvider相同，可以在任何应用程序中通过获取Contex的ContentResolver对象，然后调用ContentResolver中对应的insert、delete、query、update等方法。建立关联的Uri就是ContentProvider注册时声明的"android:authorities"属性值，ContentResolver中的这些方法会将参数传递给ContentProvider中对应的方法，由ContentProvider负责数据操作，并返回结果给ContentResolver。示例代码如下：

```java
ContentResolver contentResolver = getContentResovler();
Uri uri = Uri.parse("content://cc.turbosnail.myprovider");
Cursor cursor =contentResolver.query(uri,null,null,null,null);
```

## 6.3 主动监听ContentProvider数据变化

ContentResolver通过方法调用访问ContentProvider提供的数据，在一些情况下，如果ContentProvider共享的数据发生了变化，而ContentResolver没有进行方法调用，则它将不知道数据已经变化。如果需要在ContentProvider共享的数据发生了变化时，主动发出通知并进行响应，Android提供了ContentObserver（内容观察者）可以实现这一功能。其实现流程如下：

（1）自定义ContentObserver子类，在onChange中实现事件响应操作。

```java
public class MyContentObserver extends ContentObserver{
    private Handler handler;
    public MyContentObserver(Handler handler) {
        super(handler);
        this.handler=handler;
    }
    @Override
    public void onChange(boolean selfChange) {
        super.onChange(selfChange);
        ……//处理事件的代码
        ……//通过handler回传处理结果
    }
}
```

（2）在ContentResolver所在类（Activity、Service等）中定义ContentObserver构造方法需要的Handler对象，用来接收回传信息。

```
private Handler mHandler = new Handler() {
    public void handleMessage(Message msg) {
        switch (msg.what) {
            ……//根据回传的信息，进行相应操作
        }
    }
}
```

（3）添加ContentObserver监听。示例代码如下：
MyContentObserver myContentObserver = new MyContentObserver(myHandler);//监听事件处理完成后，通过Handler返回消息
contentResolver.registerContentObserver(uri,false,myContentObserver);//第一个参数为监听的Uri，第二个参数false表示只匹配当前Uri，true表示还可以匹配当前Uri带后缀的派生子Uri，第三个参数为监听器ContentObserver对象

（4）注销ContentObserver监听。使用完成后，需要注销监听事件。示例代码如下：
contentResolver.unregisterContentObserver(myContentObserver);

（5）如果是自定义的CntentProvider，则需要在数据更改后显示发出通知，这样ContentObserver才能监听到，代码如下：
getContext().getContentResolver().notifyChange(uri,null);//第二个参数是ContentObserver监听器对象，null表示不指定具体的监听器对象

## 6.4 为列表项选择相册中的图片作为图标

**[项目案例19] 在AccountBook项目中用户添加账目时，可以打开图片库，选择一张照片作为账目列表的图标**

在个人记账本项目中，设计如下功能：当用户在添加账目时，点击添加图片的按钮打开系统相册，允许用户选择一张图片作为当前账目的显示图标，选中后在添加账目界面上显示该张图片。用户提交保存账目后，可以在账目列表上看到选择的图片成为当前账目的图标。

实现思路：在添加账目的界面上监听添加图片图标，通过Intent跳转到图片库，并在startActivityForResult中通过URI获取到返回的图片对象，将它设置为当前界面预览图片，同时通过文件操作将图片对象保存到项目的外置扩展存储中，然后将生成的图片名称（或者完整路径）保存到数据库中。打开账目列表时，根据从数据库得到的实体Bean中的图片名称进行文件操作，读取外置扩展存储中对应的图片文件，生成图片对象，设置成当前列表的显示图标。删除账目信息时，需要检查是否有自定义的图片图标，如果有，除了要删除数据库中的记录外，还需要通过文件操作删除这张图片。

（1）在MainRecordFragment.java的成员变量中添加常量和相关变量：

```java
private static final int PHOTO_REQUEST_CODE = 1;//打开相册Intent跳转的返回码
private ImageView ivPreview;//显示预览图片
private ImageButton ibPhoto;//打开相册按钮
private Bitmap bitmap;
private String picName;
private ImageUtil imageUtil;
private Context context;
```

并在onCreateView方法中添加代码初始化相关变量,对ibPhoto添加事件响应,当用户点击时,通过Intent打开图片库选择图片:

```java
context = this.getActivity();
imageUtil = new ImageUtil(context);
ivPreview = view.findViewById(R.id.iv_preview);
ibPhoto = view.findViewById(R.id.ib_photo);
ibPhoto.setOnClickListener(new View.OnClickListener() {
    @Override
    public void onClick(View v) {
        Intent intent = new Intent();
        intent.setAction(Intent.ACTION_PICK);
        intent.setData(MediaStore.Images.Media.EXTERNAL_CONTENT_URI);
        startActivityForResult(intent,PHOTO_REQUEST_CODE);
    }
});
```

(2) 在MainRecordFragment.java中添加onActivityResult方法用来接收回传的数据,生成Bitmap图片对象,这里同时处理了基于URI和Bundle两种方式回传数据的情况。然后把生成的Bitmap对象赋给预览图片视图ivPreview,并调用ImageUtil类中的saveImage方法把图片对象保存到手机的外置扩展存储中,把生成的文件名返回用来存入数据库中:

```java
@Override
public void onActivityResult(int requestCode, int resultCode, Intent data) {
    super.onActivityResult(requestCode, resultCode, data);
    switch (requestCode) {
        case PHOTO_REQUEST_CODE:
            if (resultCode == Activity.RESULT_OK) {
                Uri uri = data.getData();//通过uri的方式返回数据
                if (uri != null) {
                    try {
                        //通过uri获取到bitmap对象
```

```
                    bitmap = MediaStore.Images.Media.getBitmap(context.getContentResolver
(), uri);
                } catch (IOException e) {
                    e.printStackTrace();
                }
            } else {
                //部分手机会返回到Bundle中
                Bundle bundle= data.getExtras();
                if (bundle!= null) {
                    bitmap = bundle.getParcelable("data");
                }
            }
            ivPreview.setImageBitmap(bitmap);//显示为界面预览图
            picName = imageUtil.saveImage(bitmap);//保存为图片文件,并返回图片名称
        }
        break;
    default:
        break;
    }
}
```

(3) 在确认按钮 btnAdd 的 onClick 事件中为 CostBean 对象赋值时,增加一个赋值操作,把生成的图片名称传入并保存,此时,当调用 addCost 方法进行数据库数据存储时,picName 也被存入数据库中。

bean.setPicPath(picName);

(4) 在 RecyclerView 的适配器类 AccountItemAdapter.java 中对界面组件赋值时,如果 getPicPath 方法获取的值不为空,则将当前项的图标设置为该值对应的图片文件对象。此时需要调用 ImageUtil 中的 readImage 方法根据数据库中的图片名称去读取外置存储中对应的文件,并生成 Bitmap 对象返回。代码清单如下:

首先声明成员变量:

private ImageUtil imageUtil;

并在构造方法中进行初始化:

imageUtil=new ImageUtil(context);

然后修改图片读取的代码段如下:

```
/*通过类型判断,为ivType设置不同的图片*/
if(bean.getPicPath()!=null){
    viewHolder.ivType.setImageBitmap(imageUtil.readImage(bean.getPicPath()));
}else {
```

```java
        if (bean.getType().equals("用餐")) {
            viewHolder.ivType.setImageResource(R.drawable.ic_eat);
        } else if (bean.getType().equals("服装")) {
            viewHolder.ivType.setImageResource(R.drawable.ic_clothes);
        } else if (bean.getType().equals("交通")) {
            viewHolder.ivType.setImageResource(R.drawable.ic_traffic);
        } else if (bean.getType().equals("娱乐")) {
            viewHolder.ivType.setImageResource(R.drawable.ic_play);
        } else {
            viewHolder.ivType.setImageResource(R.drawable.ic_icon);
        }
    }
}
```

（5）删除图片。当删除一条记录时，除了删除数据库中的记录外，还应该调用ImageUtil中的deleteImage方法删除外置存储中对应的图片文件。修改适配器类AccountItemAdapter.java中的接口方法setOnCreateContextMenuListener，在costDAO.deleteCost(bean.getCostID())语句之后添加文件删除语句如下：

```java
if(bean.getPicPath()!=null){
    imageUtil.deleteImage(bean.getPicPath());
}
```

（6）文件操作的工具类ImageUtil.java代码如下：

```java
package cc.turbosnail.accountbook;
import android.content.Context;
import android.graphics.Bitmap;
import android.graphics.BitmapFactory;
import android.os.Environment;
import java.io.File;
import java.io.FileNotFoundException;
import java.io.FileOutputStream;
import java.io.IOException;
import java.text.SimpleDateFormat;
import java.util.Date;
public class ImageUtil {
    private Context context;
    public ImageUtil(Context context) {
        this.context = context;
    }
```

/*保存图片。以当前时间字符串生成自定义的图片名称,然后保存图片到外置扩展存储中,并返回这个图片名称*/

```java
public String saveImage(Bitmap bitmap) {
    String photoName = null;
    if (Environment. getExternalStorageState(). equals(Environment. MEDIA_MOUNTED)) {//判断外部设备是否可用
        SimpleDateFormat format = new SimpleDateFormat("yyyyMMdd_HHmmss");
        Date date = new Date();
        String str = format.format(date);
        photoName = "account" + "_" + str + ".jpg";
        File path = context.getExternalFilesDir("accountPhoto");
        if (!path.exists()) {
            path.mkdir();
        }
        File file = new File(path, photoName);
        FileOutputStream fos;
        try {
            fos = new FileOutputStream(file);
            bitmap.compress(Bitmap.CompressFormat.JPEG, 20, fos);//压缩并写入。第二个参数为压缩率,[0-100],0压缩100%,100不压缩
            fos.flush();
            fos.close();
        } catch (FileNotFoundException e) {
            e.printStackTrace();
        } catch (IOException e) {
            e.printStackTrace();
        }
    }
    return photoName;
}
```

/*删除图片。根据文件名称删除外置存储中的图片文件*/

```java
public boolean deleteImage(String name) {
    boolean flag = false;
    if (Environment.getExternalStorageState().equals(Environment.MEDIA_MOUNTED)) {
        File path = context.getExternalFilesDir("accountPhoto");
        File file = new File(path, name);
        if (file.exists() && file.isFile()) {
            if (file.delete()) {
                flag = true;
```

```
            }
        }
        return flag;
}
/*读取图片。根据文件名称读取保存在外置存储中的图片文件,并返回Bitmap对象*/
public Bitmap readImage(String name) {
    Bitmap bitmap = null;
    if (Environment.getExternalStorageState().equals(Environment.MEDIA_MOUNTED)) {
        String path = context.getExternalFilesDir("accountPhoto").getAbsolutePath() + "/" + name;
        bitmap = BitmapFactory.decodeFile(path);//根据文件路径生成Bitmap对象
    }
    return bitmap;
    }
}
```

该类作为工具类使用,封装了saveImage、deleteImage、readImage三个方法,saveImage方法用来把Bitmap对象转换成图片文件存储到当前应用的外置扩展存储中,并为它生成了一个基于时间的文件名。deleteImage和readImage方法基于文件名称在外置扩展存储中删除和读取对应图片文件。

(7)运行程序,在添加账目时,点击"相册"图标,此时会打开图库,选择一张图片并确认,可看到预览图,保存账目,账目列表中可以看到用户选择的列表图标。效果如图6.2所示。

图6.2 为列表项选择图片作为图标

## 6.5 为列表项拍照并添加照片作为图标

**[项目案例20]** 在 **AccountBook** 项目中用户添加账目时，可以打开系统照相机，允许用户拍照并将当前照片作为账目列表的图标

在个人记账本项目中设计如下功能：当用户在添加账目时，点击相机图标打开系统照相机，用户拍照后，当前照片作为该账目的显示图标。用户提交保存账目后，可以在账目列表上看到该图标。

**实现思路：**

在添加账目的界面上监听相机图片按钮，通过 Intent 打开相机拍照，然后在 startActivityForResult 中获取到 Bundle 中返回的照片信息，它包含了 Bitmap 类型的图片对象。根据这个对象，再调用自定义工具类 ImageUtil 中的 saveImage 方法把图片对象存储到外置扩展存储中并返回图片名称。从相机得到图片提供了与从相册选择图片不同的数据获取方式，一旦保存到当前项目的外置扩展存储中，就可以通过工具类 ImageUtil 中定义的方法进行读取、删除等操作。需要注意的是，要为打开相机定义不同于打开相册的返回码，以便在 onActivityResult 方法中进行区分。

（1）在 MainRecordFragment.java 的成员变量中添加常量和相关变量：

```java
private static final int CAMERA_REQUEST_CODE = 2;//打开相机Intent跳转的返回码
private ImageButton ibCamera;//打开相册按钮
```

在 onCreateView 方法中对 ibCamera 进行初始化，添加事件响应，当用户点击时，通过 Intent 打开相机：

```java
ibCamera = view.findViewById(R.id.ib_camera);
ibCamera.setOnClickListener(new View.OnClickListener() {
    @Override
    public void onClick(View v) {
        Intent intent = new Intent();
        intent.setAction(MediaStore.ACTION_IMAGE_CAPTURE);//打开系统相机
        startActivityForResult(intent, CAMERA_REQUEST_CODE);//用于返回图片对象
    }
});
```

（2）在 MainRecordFragment.java 的 onActivityResult 方法的 case 语句中添加一个判断，当返回码来自相机时，读取 Bundle 中的图片信息得到 Bitmap 对象，然后设置为预览图，并调用 ImageUtil 类中的 saveImage 方法保存图像，返回图像名称。

```java
case CAMERA_REQUEST_CODE:
    if (resultCode == Activity.RESULT_OK) {
```

```
    Bundle bundle= data.getExtras();/*Bundle会返回照片的缩略图信息*/
    bitmap = bundle.getParcelable("data");
    ivPreview.setImageBitmap(bitmap);//显示为界面预览图
    picName = imageUtil.saveImage(bitmap);//保存为图片文件,并返回图片名称
}
break;
```

（3）运行程序,在添加账目时点击"相机"图标,此时会打开模拟器的相机,用户拍照并确认后（模拟器默认只能在一个带动画的效果图中模拟拍照）,可以返回账目添加界面并显示预览图,保存账目后,账目列表中可以看到这张照片成为了列表图标。效果如图6.3所示。

图6.3 为列表项拍照并作为显示图标

# 第7章 背景音乐——Service 与 BroadcastReceiver

在 Android 程序中，Activity 是在界面上运行的基础组件，当项目中一些功能不需要界面，只需要在后台提供服务，甚至当前 Activity 销毁后这些服务还需继续时，Android 提供了 Service 来完成这样的功能。Service 也是 Android 四大基础组件之一，它是一个没有界面的"Activity"，Service 在后台执行，它可以长时间运行且无需用户界面，而且可以为跨组件访问数据提供支持。当需要跨组件通信时，BroadcastReceiver 可以作为媒介使用。BroadcastReceiver 也是 Android 四大基础组件之一，它基于广播的形式接收广播信息，这些广播可以开发者自定义，也可以是系统广播。广播可以在不同组件之间传递，接收方根据特定标识进行接收。

## 7.1 Service

### 7.1.1 Service 简介

Service 作为 Android 四大基础组件之一，与 Activity 一样具有自己独立的生命周期，所不同的是：Service 只在后台运行，没有用户界面，生命周期独立于 Activity 等其他基础组件。它用来提供需要在后台长期运行的服务，例如数据下载、复杂计算、背景音乐等。

Service 可以理解为独立的、没有界面的"Activity"，启动一个 Service 后，它可以独立运行完成特定任务。这和在 Activity 中启动子线程从表面看似乎一致，但是 Service 和 Thread（线程）在概念和应用场景上是不同的：

（1）Thread 是程序执行的最小单元。Service 属于"主线程"，Service 中的耗时操作也应该开启子线程 Thread 来完成。

（2）Thread 独立运行，但是启动它的组件如 Activity 被销毁后，其他 Activity 无法再获取和控制这个线程。而 Service 作为独立组件，不受启动它的 Activity 还是否存在的影响，启动后将按照自己的生命周期运行，并接受管理。

（3）Thread 只能服务于启动它的组件，不能在多个不同的 Activity 中访问同一个 Thread 对象。而 Service 是独立的，任何 Activity 都可以对它进行访问，由它提供服务，多个 Activity 可以共享一个 Service。

在程序运行过程中，由于单个 Activity 可能被销毁，那些需要一直在后台执行的任务就难以通过 Thread 完成，而 Service 不仅可以有效地解决这样的问题，还可以和其他基础组件结合使用，完成更为复杂的任务，这使得 Service 在 Android 应用开发中有着重要作用。

### 7.1.2 创建 Service

创建 Service 一般是自定义一个 Service 类的子类。示例如下：

新建一个项目，项目名为 StartServiceDemo。在项目目录 "java" → "cc.turbosnail.startservicedemo" 上点击右键，然后选择 "New" → "Service" → "Service"，在弹出框中输入自定义的类名：MyStartService，然后创建这个类，可以看到代码如下：

```java
package cc.turbosnail.startservicedemo;
import android.app.Service;
import android.content.Intent;
import android.os.IBinder;
public class MyStartService extends Service {
    public MyStartService() {
    }
    @Override
    public IBinder onBind(Intent intent) {
        //TODO: Return the communication channel to the service.
        throw new UnsupportedOperationException("Not yet implemented");
    }
}
```

此时，默认生成一个构造方法和一个 IBinder 类型的 onBind() 方法，onBind() 用于绑定 Service 时使用。

打开 AndroidManifest.xml 配置文件可以看到，MyStartService 作为一个 Service 类已经被注册到这个文件中。

```xml
<service
    android:name=".MyStartService"
    android:enabled="true"
    android:exported="true"></service>
```

### 7.1.3 启动 Service 方式一：通过 StartService() 方法启动

在 Android 基础组件 Activity 中可以通过 StartService() 方法启动一个 Service，一旦启动，这个 Service 将在后台运行，不受启动它的基础组件的影响。与 StartService() 相关的三个需要实现的方法是：

☞ onCreate()：第一次创建 Service 时执行。

☞ onStartCommand( ):通过 StartService( )方法启动 Service 时执行。

☞ onDestroy( ):通过 stopService( )方法停止 Service 时执行,停止后会被销毁。

**[例 7.1]** 通过 StartService 方法启动 Service。

(1) 在 StartServiceDemo 项目中为自定义的 MyStartService 实现 onCreate( )、onStartCommand( )、onDestroy( )方法。

```java
package cc.turbosnail.startservicedemo;
import android.app.Service;
import android.content.Intent;
import android.os.IBinder;
import android.util.Log;
public class MyStartService extends Service {
    public MyStartService() {
    }
    @Override
    public void onCreate() {
        super.onCreate();
        Log.i("StratService******", "onCreate()");
    }
    @Override
    public int onStartCommand(Intent intent, int flags, int startId) {
        Log.i("StartService******", "onStartCommand()");
        return super.onStartCommand(intent, flags, startId);
    }
    @Override
    public void onDestroy() {
        Log.i("StartServive******", "onDestroy()");
        super.onDestroy();
    }
    @Override
    public IBinder onBind(Intent intent) {
        //TODO: Return the communication channel to the service.
        throw new UnsupportedOperationException("Not yet implemented");
    }
}
```

(2) 在 MainActivity 中通过 StartService( )方法启动 MyStartService。在 activity_main.xml 中放置两个按钮(Button),id 分别为 btn_start 和 btn_stop,在 MainActivity.java 中为按钮添加点击事件,启动和停止 Service 的方法分别为:

```
Intent intent = new Intent(MainActivity.this，MyService.class)；
startService(intent)；//在 Activity 中启动 Service
Intent intent = new Intent(MainActivity.this，MyService.class)；
stopService(intent)；//在 Activity 中停止 Service
```
代码清单如下：

```java
package cc.turbosnail.startservicedemo;
import android.content.Intent;
import android.support.v7.app.AppCompatActivity;
import android.os.Bundle;
import android.view.View;
import android.widget.Button;
public class MainActivity extends AppCompatActivity {
    private Button start, stop;
    @Override
    protected void onCreate(Bundle savedInstanceState) {
        super.onCreate(savedInstanceState);
        setContentView(R.layout.activity_main);
        start = findViewById(R.id.btn_start);
        stop = findViewById(R.id.btn_stop);
        start.setOnClickListener(new View.OnClickListener() {
            @Override
            public void onClick(View v) {
                Intent intent = new Intent(MainActivity.this, MyStartService.class);
                startService(intent);
            }
        });
        stop.setOnClickListener(new View.OnClickListener() {
            @Override
            public void onClick(View v) {
                Intent intent = new Intent(MainActivity.this, MyStartService.class);
                stopService(intent);
            }
        });
    }
}
```

运行程序，点击"启动"按钮，可以在 Logcat 窗口中看到如图 7.1 所示效果。

```
cc.turbosnail.startservicedemo I/StratService******: onCreate()
cc.turbosnail.startservicedemo I/StartService******: onStartCommand()
```

图 7.1　通过 startService 启动 Service

此时,onCreate()、onStartCommand()方法依次被调用,多次点击"启动",可以看到如图7.2所示效果。

```
cc.turbosnail.startservicedemo I/StratService******: onCreate()
cc.turbosnail.startservicedemo I/StartService******: onStartCommand()
cc.turbosnail.startservicedemo I/StartService******: onStartCommand()
cc.turbosnail.startservicedemo I/StartService******: onStartCommand()
```

图 7.2　多次执行 onStartCommand()方法

onStartCommand()被重复调用执行,但是 onCreate()方法不会重复执行。根据这个特点,可以在程序设计中把相关的业务功能代码放置在对应的方法中。点击"停止"按钮时,onDestroy()方法会被调用执行,此时 Service 终止,如图7.3所示。

```
cc.turbosnail.startservicedemo I/StartServive******: onDestroy()
```

图 7.3　停止 Service

### 7.1.4　启动 Service 方式二:通过 bindService()方法启动

除了通过 startService()方法启动 Service 之外,Android 还提供了通过 bindService()方法启动的方式,这种方式启动的 Service 与启动它的基础组件(如 Activity)有关联,基础组件终止退出时 Service 也终止,这种关联就是基于绑定模式的 Service。由于绑定关系的存在,当基础组件和 Service 之间需要进行数据交换时,可以通过绑定模式的 Service 来实现。

与 bindService()相关的四个需要实现的方法是:

☞ onCreate():第一次创建 Service 时执行。

☞ onBind():通过 bindService()方法启动 Service 时执行。

☞ onUnbind():通过 unBindService()方法断开 Service 绑定时执行。

☞ onDestroy():Service 被销毁时执行。在绑定模式下,解开绑定后,系统会销毁当前 Service。

在基础组件中启动和停止 Service 的方法声明如下:

☞ public abstract boolean bindService(Intent service, ServiceConnection conn, int flags)

☞ public abstract void unbindService(ServiceConnection conn)

这两个方法中都需要一个 ServiceConnection 类型的参数,所以必须首先构造一个 ServiceConnection 类型的对象。ServiceConnection 是一个接口,在程序运行时,当 service 创建完毕之后,会通过回调 ServiceConnection 的方法来通知关联的基础组件(如 Activity),通过这种异步的方式确保 service 的生命周期方法不会被阻塞。它的接口方法如下:

☞ public void onServiceConnected(ComponentName name, IBinder service){}

☞ public void onServiceDisconnected(ComponentName name){}

onServiceConnected()方法在Service创建后被调用,其中,如果自定义Service中的onBind()方法返回值为null,则onServiceConnected()方法将不会被调用。只有onBind()方法返回值不为null时,onServiceConnected()才会被调用,此时,方法执行的顺序为onCreate()→onBind()→onServiceConnected()。onServiceDisconnected()方法只在出现异常时才会被调用,正常退出Service时不会被调用。

在开发时,可以在启动Service的基础组件中通过内部类的形式来实现ServiceConnection接口,然后创建对象,为bindService()和unbindService()方法提供参数。

另外,可以看到自定义Service中的onBind()方法类型和ServiceConnection实现中的onServiceConnected()方法第二个参数都是IBinder类型,IBinder也是一个接口,它可以用来实现跨进程通信。在Service中,它可以用来实现Service与关联的基础组件之间的通信。在开发中,一般通过继承的方式实现这个接口,可以在自定义的Service中创建一个内部类如下:

public class MyBinder extends Binder{//定义方法,获取Service中的值}

这样,在有了ServiceConnection对象的基础组件中,就可以通过onServiceConnected()方法的第二个参数,得到自定义Service中的Binder对象,然后根据应用程序需要,调用Binder对象中的方法,以实现从Service中取值的操作。

【例7.2】通过bindService()方法启动Service,并实现Service和Activity之间的通信。

(1)新建一个项目,项目名为BindServiceDemo。在项目目录"java"→"cc.turbosnail.bindservicedemo"上点击右键,然后选择"New"→"Service"→"Service",在弹出框中输入自定义的类名:MyBindService,实现代码如下:

```
package cc.turbosnail.bindservicedemo;
import android.app.Service;
import android.content.Intent;
import android.os.Binder;
import android.os.IBinder;
import android.util.Log;
import java.util.Timer;
import java.util.TimerTask;
public class MyBindService extends Service {
    private MyBinder myBinder = new MyBinder();//内部类,用来传值
    private int num = 0;
    private boolean isRun = true;
    @Override
    public IBinder onBind(Intent intent) {
        Log.i("MyBindService******", "onBind()");
        return myBinder;
    }
//在Service中启动一个计时器线程,定时产生数字。Service启动后,获取这个数字,实现通信
    @Override
```

```java
public void onCreate() {
    Log.i("MyBindService******", "onCreate()");
    Timer timer = new Timer();
    TimerTask timerTask = new TimerTask() {
        @Override
        public void run() {
            while (isRun) {
                num++;
                try {
                    Thread.sleep(1000);
                } catch (InterruptedException e) {
                    e.printStackTrace();
                }
            }
        }
    };
    timer.schedule(timerTask, 0);
    super.onCreate();
}
@Override
public boolean onUnbind(Intent intent) {
    Log.i("MyBindService******", "onUnbind()");
    isRun = false;//终止计时器线程
    return super.onUnbind(intent);
}
@Override
public void onDestroy() {
    Log.i("MyBindService******", "onDestroy()");
    super.onDestroy();
}
/*Binder内部类,定义传值的方法*/
public class MyBinder extends Binder {
    //获取Service中的值
    public int getNum() {
        return num;
    }
}
}
```

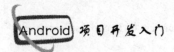

（2）在 MainActivity 中启动绑定的 Service、获取 Service 中的值、停止 Service。首先在 activity_main.xml 布局中放置三个按钮（Button），id 分别为 btn_start、btn_stop 和 btn_getValue。然后在 MainActivity 中对三个按钮进行事件监听,实现对应功能如下：

```java
package cc.turbosnail.bindservicedemo;
import android.app.Service;
import android.content.ComponentName;
import android.content.Intent;
import android.content.ServiceConnection;
import android.os.IBinder;
import android.support.v7.app.AppCompatActivity;
import android.os.Bundle;
import android.util.Log;
import android.view.View;
import android.widget.Button;
public class MainActivity extends AppCompatActivity {
    private Button btStart, btStop, btGetValue;
    private MyConnection myConnection;
    private MyBindService.MyBinder myBinder;
    @Override
    protected void onCreate(Bundle savedInstanceState) {
        super.onCreate(savedInstanceState);
        setContentView(R.layout.activity_main);
        myConnection = new MyConnection();//自定义回调接口
        btStart = findViewById(R.id.btn_start);
        btStop = findViewById(R.id.btn_stop);
        btGetValue = findViewById(R.id.btn_getValue);
        btStart.setOnClickListener(new View.OnClickListener() {
            @Override
            public void onClick(View v) {
                //启动 Service
                Intent intent = new Intent(MainActivity.this, MyBindService.class);
                bindService(intent, myConnection, Service.BIND_AUTO_CREATE);
            }
        });
        btStop.setOnClickListener(new View.OnClickListener() {
            @Override
            public void onClick(View v) {
                //解除绑定,停止 Service
```

```java
            unbindService(myConnection);
        }
    });
    btGetValue.setOnClickListener(new View.OnClickListener() {
        @Override
        public void onClick(View v) {
            //通过调用myBinder方法获取Service中的值到Activity中来实现通信
            Log.i("从service取值******", "Num:" + myBinder.getNum());
        }
    });
}
private class MyConnection implements ServiceConnection {
    @Override
    public void onServiceConnected(ComponentName name, IBinder service) {
        myBinder = (MyBindService.MyBinder) service;
        Log.i("MyBindService******", "ServiceConnected");
    }
    @Override
    public void onServiceDisconnected(ComponentName name) {
        Log.i("MyBindService******", "ServiceDisconnected");
    }
}
}
```

运行程序，通过按钮启动Service后，Service中的计时器启动，每秒钟num的值会自增，此时，点击获取值的按钮，可以在日志中看到num变化后的值，即myBinder.getNum()方法将Serv中的值读取到了Activity中来，实现了它们之间的传值功能。在Logcat窗口中可以看到如图7.4所示效果。

```
cc.turbosnail.bindservicedemo   I/MyBindService******: onCreate()
cc.turbosnail.bindservicedemo   I/MyBindService******: onBind()
cc.turbosnail.bindservicedemo   I/MyBindService******: ServiceConnected
cc.turbosnail.bindservicedemo   I/从service取值******: Num: 14
cc.turbosnail.bindservicedemo   I/从service取值******: Num: 15
cc.turbosnail.bindservicedemo   I/从service取值******: Num: 17
cc.turbosnail.bindservicedemo   I/MyBindService******: onUnbind()
cc.turbosnail.bindservicedemo   I/MyBindService******: onDestroy()
```

图7.4 bindService()方法启动Service

## 7.2 音乐播放器

MediaPlayer 是 Android 多媒体框架（Media Framework）中的一个重要类，通过它可以播放音频和视频多媒体文件。MediaPlayer 提供了加载数据源（setDataSource）、准备（prepare）、开始（start）、暂停（pause）、停止（stop）、重置（reset）、释放资源（release）等方法来完成播放流程，需要注意的是每一次开始（start）播放前，必须先进行准备（prepare）操作。MediaPlayer 还提供了设置循环（setLooping）、指定播放的位置（seekTo）等方法，并能结合示波器、重低音控制器、噪声压制器等相关类来实现播放效果管理的功能。

（1）播放项目目录中的音频资源文件。

在 res 目录中新建 raw 资源目录。在 res 目录上点击右键，选择"New"→"Folder"→"Raw Resources Folder"，确认后，可以看到 res/raw 目录。

把音乐文件复制到 raw 目录中，然后在 Java 代码中通过 MediaPlayer 播放。此时，可以通过 create() 来初始化对象，这种方式已经默认进入准备状态，创建好对象后直接通过 start 播放即可。MediaPlayer 的 create 方法如下：

- public static MediaPlayer create(Context context, int resid)，通过给定的 Id 来创建一个 MediaPlayer 实例。
- public static MediaPlayer create(Context context, Uri uri)，通过给定的 Uri 来创建一个 MediaPlayer 实例。

示例代码如下：

MediaPlayer mediaPlayer = MediaPlayer.create(this, R.raw.music_bg);
mediaPlayer.strat();

可以通过以下方法对播放过程进行控制：

- mediaPlayer.setLooping(true);//设置是否循环播放。
- mediaPlayer.setVolume(1.0F, 1.0F);//设置音量，左声道和右声道，float 型，[0f, 1f] 之间。
- mediaPlayer.seekTo(5000);//跳转到对应时间进行播放，int 型，毫秒数。
- mediaPlayer.pause();//暂停。
- mediaPlayer.stop();//停止。
- mediaPlayer.reset();//重置，回到 Idle 状态。
- mediaPlayer.release();//播放结束后释放资源。
- Boolean isPlay = mPlayer.isPlaying();//判断是否正在播放。
- int currentPosition = mediaPlayer.getCurrentPosition();//获取播放进度，int 型，毫秒数。

由于流媒体数据比较占系统资源，在音乐播放结束后一定要通过 release() 方法释放资源，示例代码如下：

if (mediaPlayer != null && mediaPlayer.isPlaying()) {

```
        mediaPlayer.stop();
        mediaPlayer.release();
        mediaPlayer = null;
}
```

（2）播放来自存储卡、网络等位置的音频文件。

当音频文件来自存储卡、网络等位置时，首先通过 MediaPlayer 的构造方法初始化播放器对象，然后通过 setDataSource() 方法设置数据源，接着按照准备、开始播放的顺序播放音乐。MediaPlayer 只有一个无参数的构造方法 MediaPlayer()，setDataSource() 方法根据数据源不同，常用的有如下几个：

☞ public void setDataSource(String path)，通过一个路径字符串设置数据源，可以是本地路径，也可以是网络路径。

☞ public void setDataSource(Context context, Uri uri)，通过 Uri 设置数据源。

示例代码如下：

```
MediaPlayer mediaPlayer = new MediaPlayer();
mediaPlayer.setDataSource("/sdcard/test.mp3");
mediaPlayer.prepare();
mediaPlayer.start();
```

播放过程控制的相关方法与播放项目目录中的音频资源文件相同，需要说明的是，prepare() 方法是以同步的方式装载资源，可能会造成 UI 界面的卡顿甚至程序无响应 ANR（Application Not Responding）错误，Android 还提供了一个异步的方法 prepareAsync() 来准备资源，这个方法通过 setOnPreparedListener() 设置监听器对装载完成事件进行监听，装载资源完成之后回调。推荐通过异步的方式装载资源，示例代码如下：

```
mediaPlayer.prepareAsync();
mediaPlayer.setOnPreparedListener(new OnPreparedListener() {
    @Override
    public void onPrepared(MediaPlayer mp) {
        mediaPlayer.start();
    }
});
```

（3）播放过程中的事件监听。

除了对异步装载资源可以进行事件监听之外，MediaPlayer 还提供了一些与流媒体播放相关的事件监听，例如：

① 监听播放结束事件。播放结束时触发这个事件，可以将后续任务放在这里处理，如重播、播放下一首等，示例代码如下：

```
mediaPlayer.setOnCompletionListener(new OnCompletionListener() {
```

```java
    @Override
    public void onCompletion(MediaPlayer mp) {
        ……//事件响应的代码,如加载另一首,然后播放
    }
});
```

② 监听播放异常事件。在播放过程中出现解析或者源文件中有错误等问题时,将会严重影响程序运行,这种情况可以通过监听播放异常事件来处理,以保证程序的健壮性。当出现这个事件时,根据监视器中的方法来响应,如重置、跳转到下一首等。

```java
mediaPlayer.setOnErrorListener(new OnErrorListener() {
    @Override
    public boolean onError(MediaPlayer mp, int what, int extra) {
        ……//事件响应的代码,如重置
    }
});
```

③ 监听定位完成事件。seekTo(int)方法是异步的,它返回后实际的定位操作可能还需要一段时间才能完成,为了保证定位完成之后再进行后续操作,可以通过监听定位完成事件来处理。

```java
mediaPlayer.setOnSeekCompleteListener(new OnSeekCompleteListener(){
    @Override
    public void onSeekComplete (MediaPlayer mp){
        ……//事件响应的代码,如开始播放
    }
});
```

**[例7.3]** 设计并实现一款简单的音乐播放器。

（1）新建一个项目,项目名为MediaPlayerDemo。

（2）在res目录上点击右键,选择"New"→"Folder"→"Raw Resources Folder",创建raw目录。然后把音频文件"music_bg.mp3"复制到这个目录中。

（3）把制作好的"播放""暂停""重播""停止"等图标复制到drawable目录中。

（4）封装音乐播放、暂停、重播、停止等几个方法,然后在播放器对应的界面按钮事件中调用,代码清单如下:

```java
package cc.turbosnail.mediaplayerdemo;
import android.media.MediaPlayer;
import android.os.Bundle;
import android.view.View;
import android.widget.ImageButton;
```

```java
import android.widget.SeekBar;
import android.support.v7.app.AppCompatActivity;
public class MainActivity extends AppCompatActivity {
    private ImageButton ibtnStart, ibtnStop, ibtnPause, ibtnReplay;//播放器按钮
    private SeekBar seekBar;//进度条
    private MediaPlayer mediaPlayer = null;
    private Object obj = new Object();//对象锁,播放线程暂停时,让进度条线程进入等待状态
    private Thread seekThread;
    private boolean isRun = false;//进度条线程控制
    private boolean suspended = false;//进度条线程等待状态控制
    @Override
    protected void onCreate(Bundle savedInstanceState) {
        super.onCreate(savedInstanceState);
        setContentView(R.layout.activity_main);
        ibtnStart = findViewById(R.id.ibtn_start);
        ibtnStop = findViewById(R.id.ibtn_stop);
        ibtnPause = findViewById(R.id.ibtn_pause);
        ibtnReplay = findViewById(R.id.ibtn_replay);
        seekBar = findViewById(R.id.seeBar);
        ibtnStart.setOnClickListener(new View.OnClickListener() {
            @Override
            public void onClick(View v) {
                if (mediaPlayer == null) {
                    play();
                } else {
                    if (!mediaPlayer.isPlaying()) { //如果处于暂停状态
                        int progress = seekBar.getProgress();
                        int currentPosition = mediaPlayer.getCurrentPosition();
                        continuePlay(progress, currentPosition);
                    }
                }
            }
        });
        ibtnPause.setOnClickListener(new View.OnClickListener() {
            @Override
            public void onClick(View v) {
                pause();
            }
        });
```

```java
        ibtnReplay.setOnClickListener(new View.OnClickListener() {
            @Override
            public void onClick(View v) {
                if (mediaPlayer == null) { //播放器对象还未创建或者已经销毁
                    play();
                } else {
                    if (!mediaPlayer.isPlaying()) { //暂停状态
                        continuePlay(0, 0);
                    } else { //正在播放状态不需要唤醒线程的操作
                        mediaPlayer.seekTo(0);
                        mediaPlayer.start();
                    }
                }
            }
        });
        ibtnStop.setOnClickListener(new View.OnClickListener() {
            @Override
            public void onClick(View v) {
                stop();
            }
        });
        seekBar.setOnSeekBarChangeListener(new SeekBar.OnSeekBarChangeListener() {
            @Override
            public void onStartTrackingTouch(SeekBar seekBar) {
                pause();
            }
            @Override
            public void onStopTrackingTouch(SeekBar seekBar) {
                if (mediaPlayer != null && !mediaPlayer.isPlaying()) {
                    int progress2 = seekBar.getProgress();
                    int currentPosition2 = mediaPlayer.getDuration() * progress2 / 100;
                    continuePlay(progress2, currentPosition2);
                }
            }
            @Override
            public void onProgressChanged(SeekBar seekBar, int progress, boolean fromUser) {
            }
        });
    }
```

```java
/*线程用来根据音乐播放进度绘制进度条*/
class SeekThread extends Thread {
    int duration = mediaPlayer.getDuration();//当前音乐总长度
    int currentPosition = 0;
    public void run() {
        while (isRun) {
            currentPosition = mediaPlayer.getCurrentPosition();//音乐当前播放位置
            seekBar.setProgress(currentPosition * 100 / duration);
            try {
                Thread.sleep(300);
            } catch (InterruptedException e) {
                e.printStackTrace();
            }
            synchronized (obj) {
                while (suspended) {
                    try {
                        obj.wait(); //音乐暂停时让进度条线程也暂停
                    } catch (InterruptedException e) {
                        e.printStackTrace();
                    }
                }
            }
        }
    }
}
/*初始化播放,一个是音乐播放,一个是线程控制的进度条绘制*/
public void play() {
    mediaPlayer = MediaPlayer.create(MainActivity.this, R.raw.music_bg);
    mediaPlayer.setOnCompletionListener(new MediaPlayer.OnCompletionListener() {
        @Override
        public void onCompletion(MediaPlayer mp) {
            if (mediaPlayer != null) {
                stop();
            }
        }
    });
    isRun = true;
    seekThread = new SeekThread();
    mediaPlayer.start();
```

```
            seekThread.start();
        }
        public void stop() {
            if (mediaPlayer != null) {
                seekBar.setProgress(0);
                isRun = false;
                mediaPlayer.stop();
                mediaPlayer.release();
                mediaPlayer = null;
                seekThread = null;
            }
        }
        public void pause() {
            if (mediaPlayer != null && mediaPlayer.isPlaying()) {
                mediaPlayer.pause();
                suspended = true;
            }
        }
        /*在从暂停状态恢复播放时使用,除了继续播放音乐外,还需要唤醒等待中的进度条绘制
线程*/
        public void continuePlay(int progress, int currentPosition) {
            mediaPlayer.seekTo(currentPosition);
            mediaPlayer.start();
            seekBar.setProgress(progress);
            suspended = false;
            synchronized (obj) {
                obj.notify();//唤醒线程,开始绘制进度条
            }
        }
    }
}
```

运行程序,播放音乐时,进度条会前进,运行效果如图7.5所示。

图 7.5 音乐播放器

注意,在本例中启动了一个用来绘制进度条的线程 SeekThread,不断获取当前音乐播放的进度,根据进度比例绘制进度条,当音乐播放暂停时,由对象锁让进度条线程进入等待状态,再次播放时唤醒线程,让进度条继续绘制。

## 7.3 BroadcastReceiver

### 7.3.1 BroadcastReceiver 简介

当在 Activity 中启动 Service 后,Activity 和 Service 在运行过程中往往需要进行数据传递,例如由 Activity 控制 Service 中的音乐播放器是否播放、Service 定时向 Activity 传递数据下载情况以便 Activity 更新 UI 界面等,此时,通过 BroadcastReceiver 可以比较方便地实现不同基础组件之间的数据传递工作。

BroadcastReceiver 作为 Android 四大基础组件之一,可以方便地实现不同基础组件、不同应用程序进程之间的通信,它可以监听全局的广播信息,这些广播信息可以由开发者自行开发的程序发出,也可以由 Android 系统程序发出,BroadcastReceiver 可以监听并接收这些广播信息。

### 7.3.2 发送广播信息

广播信息以 Intent 的形式由 Contex 的 sendBroadcast 或者 sendOrderedBroadcast 等方法进行发送,这意味着开发者可以把数据封装在 Intent 中,并在任何时候都可以发出广播信息。同时,在发送广播信息时可以设定 Intent 的 Action 属性,这样只有能匹配的接收方才可以接收这个广播信息,实现定向广播的功能。

(1)发送普通广播。

普通广播的广播信息是所有匹配的接收者都可以接收的,接收者不能修改广播信息,也不能把广播信息按次序转给下一个接收者,即所有接收者都可以同时接收这个广播信息。发送普通广播的方法为 public void sendBroadcast(Intent intent),示例代码如下:

```
Intent intent = new Intent();
intent.setAction("cc.turbosnail.NormalBroadcastDemo");//接收方匹配Action才能接收广播
intent.putExtra("data","普通广播信息");
sendBroadcast(intent);
```

（2）发送有序广播。

有序广播与普通广播的区别是，有序广播的接收者会按优先级接收广播信息，优先级高的接收者先接收，然后优先级低的接收者再接收，顺次接收。优先级高的接收者可以往广播信息中添加信息，然后一并传给优先级低的接收者。同时，优先级高的接收者还有权终止广播信息传播，一旦终止，优先级低的接收者将不能再接收这个广播信息。发送有序广播的方法为public void sendOrderedBroadcast（Intent intent，String receiverPermission），receiverPermission是对接收者的权限声明，null表示不需要任何权限。示例代码如下：

```
Intent intent = new Intent();
intent.setAction("cc.turbosnail.OrderedBroadcastDemo");//接收方匹配Action才能接收广播
intent.putExtra("data","有序广播信息");
sendOrderedBroadcast (intent,null);
```

### 7.3.3 接收广播信息

广播接收器BroadcastReceiver对象用来监听和接收广播信息，接收到相应广播后，会自动回调onReceive()方法，因此，接收广播信息后的处理、与其他组件的交互操作一般都可以放在onReceive()方法中。

（1）创建BroadcastReceiver子类作为广播接收器。

```
public class MyBroadcastReceiver extends BroadcastReceiver {
@Override
    public void onReceive(Context context, Intent intent) {
        String msg = intent.getStringExtra("data");
        ……//写入接收广播信息后的操作
    }
}
```

对有序广播来说，优先级高的接收者可以在onReceive()方法中调用如下方法终止广播继续向下传播：

abortBroadcast();

也可以修改获取到的值，通过调用setResultExtras()方法往下传递修改后的值，示例代码如下：

```
Bundle bundle = new Bundle();
```

```
bundle.putString("data", data + "追加的数据");
setResultExtras(bundle);
```

（2）注册广播接收器。

① 在AndroidManifest.xml里通过<receive>标签注册自定义的广播接收器。

```
<receiver android:name=".MyBroadcastReceiver">
    <intent-filter>
    <action android:name="android.net.conn.CONNECTIVITY_CHANGE"/>
    </intent-filter>
</receiver>
```

对有序广播来说，需要在<intent-filter>标签中添加优先级属性说明，优先级的取值范围为[-1000,1000]，数值越大，优先级越高。具体写法如下：

```
<intent-filter android:priority="900">
```

② 在代码中调用Context.registerReceiver()方法注册自定义的广播接收器。

```
MyBroadcastReceiver myBroadcastReceiver = new MyBroadcastReceiver();
IntentFilter intentFilter = new IntentFilter();
intentFilter.addAction("cc.turbosnail.NormalBroadcastDemo");
registerReceiver(myBroadcastReceiver, intentFilter);
```

注意：在代码中注册广播接收器后，要在使用完成后取消注册，代码如下：

```
unregisterReceiver(myBroadcastReceiver);
```

一般情况下，建议在onResume()方法中注册广播接收器，在onPause()方法中取消注册。

**[例7.4]** 在一个Service中发送广播，并在一个Activity中进行接收。

（1）新建项目，项目名称为BroadcastDemo。

（2）新建一个Activity，类名为ReceiverActivity，同时，新建一个Service类，类名为WorkService。项目目录如图7.6所示。

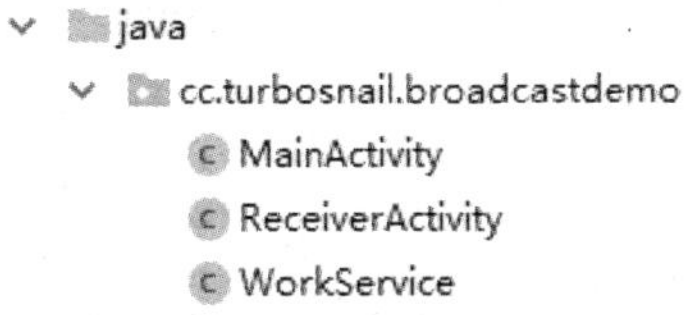

```
v ■ java
    v ■ cc.turbosnail.broadcastdemo
        © MainActivity
        © ReceiverActivity
        © WorkService
```

图7.6 BroadcastDemo项目目录

（3）在Service中播放广播。在WorkService中启动一个线程，定时播放广播信息，广播信息为不断递增的一个数字。代码清单如下：

```
package cc.turbosnail.broadcastdemo;
import android.app.Service;
import android.content.Intent;
import android.os.IBinder;
public class WorkService extends Service {
```

```java
    private boolean isRun = true;
    @Override
    public IBinder onBind(Intent intent) {
        // TODO: Return the communication channel to the service.
        throw new UnsupportedOperationException("Not yet implemented");
    }
    @Override
    public int onStartCommand(Intent intent, int flags, int startId) {
        return super.onStartCommand(intent, flags, startId);
    }
    @Override
    public void onDestroy() {
        isRun = false;//终止线程,不再发送广播消息
        super.onDestroy();
    }
    @Override
    public void onCreate() {
        new Thread(new Runnable() {
            @Override
            public void run() {
                int count = 0;
                while (isRun) {
                    Intent intent = new Intent();
                    intent.setAction("cc.turbosnail.broadcastdemo");
                    count++;
                    intent.putExtra("data", "广播信息" + count);
                    sendBroadcast(intent);
                    try {
                        Thread.sleep(3000);
                    } catch (InterruptedException e) {
                        e.printStackTrace();
                    }
                }
            }
        }).start();
        super.onCreate();
    }
}
```

（4）接收广播。在ReceiverActivity中创建内部类MyReceiver，并注册用来接收广播信息，接收的广播信息通过Toast进行提示显示。代码清单如下：

```java
package cc.turbosnail.broadcastdemo;
import android.content.BroadcastReceiver;
import android.content.Context;
import android.content.Intent;
import android.content.IntentFilter;
import android.support.v7.app.AppCompatActivity;
import android.os.Bundle;
import android.widget.Toast;
public class ReceiverActivity extends AppCompatActivity {
    MyReceiver myReceiver = new MyReceiver();
    @Override
    protected void onCreate(Bundle savedInstanceState) {
        super.onCreate(savedInstanceState);
        setContentView(R.layout.activity_receiver);
    }
    @Override
    protected void onResume() {
        super.onResume();
        IntentFilter intentFilter = new IntentFilter();
        intentFilter.addAction("cc.turbosnail.broadcastdemo");
        registerReceiver(myReceiver, intentFilter);
    }
    @Override
    protected void onPause() {
        super.onPause();
        unregisterReceiver(myReceiver);
    }
    public class MyReceiver extends BroadcastReceiver {
        @Override
        public void onReceive(Context context, Intent intent) {
            String data = intent.getStringExtra("data");
            Toast.makeText(ReceiverActivity.this, data, Toast.LENGTH_LONG).show();
        }
    }
}
```

（5）在MainActivity中启动和停止Service。界面上放置三个按钮，分别用来启动服务、停止服务、跳转到ReceiverActivity界面上，代码清单如下：

```java
package cc.turbosnail.broadcastdemo;
import android.content.Intent;
import android.support.v7.app.AppCompatActivity;
import android.os.Bundle;
import android.view.View;
import android.widget.Button;
import android.widget.Toast;
public class MainActivity extends AppCompatActivity {
    private Button btnStart, btnReceiver, btnStop;
    @Override
    protected void onCreate(Bundle savedInstanceState) {
        super.onCreate(savedInstanceState);
        setContentView(R.layout.activity_main);
        btnStart = findViewById(R.id.btn_start);
        btnReceiver = findViewById(R.id.btn_receiver);
        btnStop = findViewById(R.id.btn_stop);
        btnStart.setOnClickListener(new View.OnClickListener() {
            @Override
            public void onClick(View v) {
                Intent intent = new Intent(MainActivity.this, WorkService.class);
                startService(intent);
                Toast.makeText(MainActivity.this, "Service 启动，开始广播", Toast.LENGTH_LONG).show();
            }
        });
        btnReceiver.setOnClickListener(new View.OnClickListener() {
            @Override
            public void onClick(View v) {
                Intent intent = new Intent(MainActivity.this, ReceiverActivity.class);
                startActivity(intent);
            }
        });
        btnStop.setOnClickListener(new View.OnClickListener() {
            @Override
            public void onClick(View v) {
                Intent intent = new Intent(MainActivity.this, WorkService.class);
```

## 第 7 章 背景音乐——Service 与 BroadcastReceiver

```
            stopService(intent);
            Toast.makeText(MainActivity.this, "Service 停止，停止广播", Toast.LENGTH_LONG).show();
        }
    });
    }
}
```

运行程序，如图 7.7 所示，在主界面上点击"启动 SERVICE 定时发送广播"按钮，然后点击"跳转到广播接收界面"，在新的界面上可以看到，每隔 3 秒，当前界面收到广播信息，并通过 Toast 进行提醒。

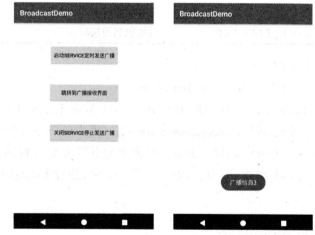

图 7.7　发送与接收广播信息

### 7.3.4　接收系统广播

Android 内置了大量系统广播，在相应事件发生时都会发出广播，开发者可以直接接收这些广播信息，并以此来实现丰富的软件功能，如网络断开时提醒、电量提醒、锁屏桌面、开机启动、短信拦截等。常见的系统广播及标识 Action 常量如表 7.1 所示。

表 7.1　系统广播常量列表

| Action 常量 | 广播事件 |
| --- | --- |
| Intent.ACTION_SCREEN_OFF | 屏幕关闭 |
| Intent.ACTION_SCREEN_ON | 屏幕打开 |
| Intent.ACTION_CLOSE_SYSTEM_DIALOGS | 屏幕锁屏 |
| Intent.ACTION_BATTERY_LOW | 电量不足 |
| Intent.ACTION_BATTERY_OKAY | 电量充满时 |
| Intent.ACTION_BATTERY_CHANGED | 电量变化 |
| Intent.ACTION_SHUTDOWN | 关闭系统 |

续表

| Action 常量 | 广播事件 |
|---|---|
| Intent.ACTION_REBOOT | 重启设备 |
| Intent.ACTION_BOOT_COMPLETED | 系统启动完成后 |
| Intent.ACTION_HEADSET_PLUG | 插入耳机 |
| Intent.ACTION_NEW_OUTGOING_CALL | 拨打电话 |
| android.net.conn.CONNECTIVITY_CHANGE | 网络变化 |
| Intent.ACTION_CAMERA_BUTTON | 拍照按键按下 |
| Intent.ACTION_MEDIA_CHECKING | 插入SD卡等外部储存装置 |
| Intent.ACTION_PACKAGE_ADDED | APK安装完成 |
| Intent.ACTION_PACKAGE_REMOVED | APK卸载完成 |
| android.provider.Telephony.SMS_RECEIVED | 系统收到短信时 |

**【例7.5】** 手机电量提醒。

（1）新建一个项目，项目名为BatteryToastDemo。

（2）在activity_main.xml布局文件中放置一个文本标签，id设置为tv_battery。在MainActivity中创建一个内部类BatteryBroadcastReciver用来监听电量变化的广播信息，常量为Intent.ACTION_BATTERY_CHANGED。根据当前电量和总电量的比例关系，计算电量百分比输出显示到文本标签中，同时，当电量低于10%时，通过Toast进行提示。代码清单如下：

```java
package cc.turbosnail.batterytoastdemo;
import android.content.BroadcastReceiver;
import android.content.Context;
import android.content.Intent;
import android.content.IntentFilter;
import android.support.v7.app.AppCompatActivity;
import android.os.Bundle;
import android.widget.TextView;
import android.widget.Toast;
public class MainActivity extends AppCompatActivity {
    BatteryBroadcastReciver batteryBroadcastReciver;
    TextView tvBattery;
    @Override
    protected void onCreate(Bundle savedInstanceState) {
        super.onCreate(savedInstanceState);
        setContentView(R.layout.activity_main);
        tvBattery = findViewById(R.id.tv_battery);
```

```java
    }
    @Override
    protected void onResume() {
        super.onResume();
        batteryBroadcastReciver = new BatteryBroadcastReciver();
        IntentFilter intentFilter = new IntentFilter(Intent.ACTION_BATTERY_CHANGED);
        registerReceiver(batteryBroadcastReciver, intentFilter);
    }
    @Override
    protected void onPause() {
        super.onPause();
        unregisterReceiver(batteryBroadcastReciver);
    }
    public class BatteryBroadcastReciver extends BroadcastReceiver {
        @Override
        public void onReceive(Context context, Intent intent) {
            if (intent.getAction().equals(Intent.ACTION_BATTERY_CHANGED)) {
                //得到当前电量
                int level = intent.getIntExtra("level", 0);
                //取得总电量
                int total = intent.getIntExtra("scale", 100);
                tvBattery.setText("当前电量:" + (level * 100) / total + "%");
                //当电量小于10%时进行提示
                if (level < 10) {
                    Toast.makeText(MainActivity.this, "当前电量已小于10%", Toast.LENGTH_LONG).show();
                }
            }
        }
    }
}
```

（3）运行程序，并打开模拟器右侧开关栏最下方"…"按钮，然后在"Battery"选项中拖动电量状态条，随着电量的变化，可以看到模拟器上电量值也在变化，当电量小于10%时，会弹出如图7.8所示提示框。

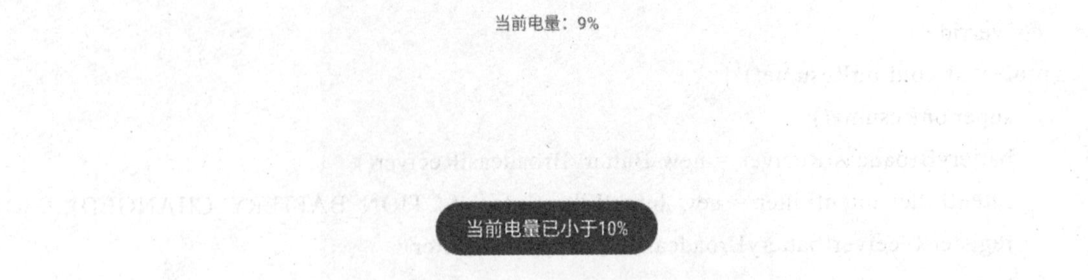

当前电量: 9%

当前电量已小于10%

图 7.8　手机电量提醒

## 7.4　背景音乐——在 Service 中播放音乐

当应用程序需要有背景音乐时，可以通过在 Service 中播放音乐的方式来实现，只要 Service 没有被终止，音乐就会一直播放。

**[项目案例21] 在 AccountBook 项目中实现一个背景音乐播放的功能**

用户在登录界面上通过复选框选择是否播放背景音乐，如果选择播放，则当用户登录成功后，自动播放背景音乐。同时，用户下一次打开登录界面时，自动显示上次的选择状态。

本例将基于 bindService 的方式启动一个 Service，在 Service 中通过 MediaPlayer 播放音乐。在登录界面 LoginActivity 中记录是否播放音乐的状态到 SharedPreferences 中，在 MainActivity 中根据 SharedPreferences 中的值来判断是否绑定和播放背景音乐。当 MainActivity 终止时，由于绑定关系，Service 会解除绑定、终止音乐播放。

（1）在 AccountBook 项目的 res 目录上点击右键，选择 "New" → "Folder" → "Raw Resources Folder"，创建 raw 目录，然后把音频文件 "music_bg.mp3" 复制到这个目录中。

（2）在 AccountBook 项目中新建一个 Service，命名为 MusicService。在这个 Service 中启动音乐播放器，并对播放器 onCompletion 事件进行监听以实现重复播放，对 onError 事件进行监听从而在出现播放错误时释放当前资源。代码清单如下：

```
package cc.turbosnail.accountbook;
import android.app.Service;
import android.content.Intent;
import android.media.MediaPlayer;
import android.os.IBinder;
import java.io.IOException;
public class MusicService extends Service {
    private MediaPlayer mediaPlayer;
    @Override
    public IBinder onBind(Intent intent) {
```

```java
            mediaPlayer.start();
            mediaPlayer.setOnCompletionListener(new MediaPlayer.OnCompletionListener() {
                @Override
                public void onCompletion(MediaPlayer mediaPlayer) {
                    //循环播放
                    try {
                        mediaPlayer.start();
                    } catch (IllegalStateException e) {
                        e.printStackTrace();
                    }
                }
            });
            //播放音乐时发生错误的事件处理
            mediaPlayer.setOnErrorListener(new MediaPlayer.OnErrorListener() {
                @Override
                public boolean onError(MediaPlayer mediaPlayer, int what, int extra) {
                    //释放资源
                    try {
                        mediaPlayer.release();
                    } catch (Exception e) {
                        e.printStackTrace();
                    }
                    return false;
                }
            });
            return null;//返回一个null
        }
    @Override
    public void onCreate() {
        super.onCreate();
        mediaPlayer = new MediaPlayer();
        mediaPlayer = MediaPlayer.create(MusicService.this, R.raw.music_bg);
        try {
            mediaPlayer.prepare();
        } catch (IllegalStateException e) {
            e.printStackTrace();
        } catch (IOException e) {
            e.printStackTrace();
        }
```

```java
    }
    @Override
    public boolean onUnbind(Intent intent) {
        //停止播放、释放资源
        mediaPlayer.stop();
        mediaPlayer.release();
        return super.onUnbind(intent);
    }
    @Override
    public void onDestroy() {
        super.onDestroy();
    }
}
```

（3）在登录 LoginActivity 中记录用户选择状态到 SharedPreferences 里面。在项目登录界面 activity_login.xml 中添加一个复选框（CheckBox），id 为 ck_music。在 LoginActivity 中获取这个复选框对象，命名为 ckMusic，然后监听这个复选框，把用户的选择状态以 boolean 类型存入 SharedPreferences 中。同时，在 LoginActivity 启动时，首先读取 SharedPreferences 中的用户选择状态值，并作为初始值赋给复选框。在 LoginActivity 初始化数据的方法 initData() 中添加如下代码：

```java
if(sp.getBoolean("isMusic",false)){
    ckMusic.setChecked(true);
}
```

在 LoginActivity 的 onCreate 方法中添加如下代码：

```java
ckMusic.setOnCheckedChangeListener(new CompoundButton.OnCheckedChangeListener() {
    @Override
    public void onCheckedChanged(CompoundButton buttonView, boolean isChecked) {
        editor = sp.edit();
        editor.putBoolean("isMusic", ckMusic.isChecked());
        editor.commit();
    }
});
```

（4）在 MainActivity 的 onCreate 方法中获取 SharedPreferences 中的值，如果该值为 true，则通过 bindService 绑定和启动 MusicService，实现背景音乐播放的功能。

```java
sp = getSharedPreferences("user", MODE_PRIVATE);
if (sp.getBoolean("isMusic", false)) {
```

```java
Intent intent = new Intent(MainActivity.this, MusicService.class);
ServiceConnection connection = new ServiceConnection() {
    @Override
    public void onServiceConnected(ComponentName name, IBinder service) {
    }
    @Override
    public void onServiceDisconnected(ComponentName name) {
    }
};
bindService(intent, connection, Service.BIND_AUTO_CREATE);
}
```

注意,由于Android对Activity生命周期进行管理时,程序从退入排队的栈中到彻底结束还有一定的时间,所以最好在程序退出的代码中手动停止服务或者手动结束整个进程。例如实现一个用户连续点击两次返回键退出程序的功能。

在MainActivity中定义两个成员变量,代码清单如下:

```java
private boolean isExit = false;
private Handler handler = new Handler() {
    @Override
    public void handleMessage(Message msg) {
        super.handleMessage(msg);
        isExit = false;
    }
};
```

然后在MainActivity中监听一个界面事件,基于Handler实现用户连续点击两次返回键退出程序的功能,立即终止整个进程的办法是,在finish方法执行后,立即调用System.exit(0)方法终止整个进程。代码清单如下:

```java
@Override
public boolean onKeyDown(int keyCode, KeyEvent event) {
    if (keyCode == KeyEvent.KEYCODE_BACK) {
        if (!isExit) {
            isExit = true;
            Toast.makeText(getApplicationContext(), "再按一次退出程序", Toast.LENGTH_SHORT).show();
            handler.sendEmptyMessageDelayed(0, 2000);
        } else {
            MainActivity.this.finish();//进入销毁的排队序列
```

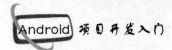

```
                System.exit(0);//终止整个进程
            }
            return false;
        }
        return super.onKeyDown(keyCode, event);
}
```

以上代码段基于Handler实现了用户连续点击两次返回键退出程序的功能,第一次点击提示"再按一次退出程序",第二次点击时结束程序,在需要对用户连续点击事件进行监听时,可以模仿这种方式去实现。

# 第8章 动画Logo——绘图与动画

以动画效果来展现欢迎界面中的Logo、界面切换时具有不同的切换效果,这些功能可以有效地提升软件的用户体验感,各种绘图、动画制作本身也是Android的一个应用方向。按钮等各种组件本质上就是在屏幕上绘制出特定样式的View。开发者可以通过图形绘制的方式自定义View,也可以直接在View中绘制图形图像,并对图像对象进行处理。动画基于图形图像绘制的基础实现,Android提供了帧动画、补间动画、属性动画等多种方式来为界面元素实现动画效果,还为Activity的切换提供了类似幻灯片的跳转动画效果。为了进一步提升动画效果、方便游戏开发,Android专门提供了SurfaceView,让开发者能更便捷地实现丰富的动画功能。

## 8.1 绘图

### 8.1.1 基本图形绘制

Android为View提供了一个绘图的方法onDraw(),重写这个方法就可以为各种View组件进行绘图操作,这个方法在组件初始化或者发生变化时会自动回调,当需要手动调用onDraw()方法时,可以通过调用View对象的invalidate()或者postinvalidate()方法,它们刷新组件时会回调onDraw()方法。绘图时Canvas类用来作为当前View组件的画板,Paint类用来作为画笔。Canvas类常用的绘图方法如下:

- drawText(String text, float x, float y, Paint paint),绘制字符串,参数text为字符串,起始位置坐标为(x,y)。
- drawPoint(float x, float y, Paint paint),绘制一个点,参数为坐标(x,y)。
- drawLine(float startX, float startY, float stopX, float stopY, Paint paint),绘制一条直线,参数分别代表起点和终点的X、Y坐标。
- drawRect(float left, float top, float right, float bottom, Paint paint),绘制一个矩形,参数为矩形的左上角坐标(left,top)和右下角坐标(right,bottom)。
- drawCircle(float cx, float cy, float radius, Paint paint),绘制一个圆,参数为圆心坐

标(cx,cy)和圆的半径radius。
- drawOval(float left, float top, float right, float bottom, Paint paint),绘制一个椭圆,前四个参数定义了椭圆外边框矩形。
- drawArc(float left, float top, float right, float bottom, float startAngle, float sweepAngle, boolean useCenter, Paint paint),绘制圆弧,前四个参数定义圆弧所在圆的外边框矩形,startAngle代表起始弧度(顺时针),sweepAngle代表绘制的弧度数(顺时针),useCenter如果为true,则表示含圆心,即构成一个扇形,如果为false,则只是弧形部分。
- drawPath(Path path, Paint paint),绘制任意多边形,参数path代表一个封闭的路径,它的定义可以如下:

```
Path path= new Path();
/*设置多边形的点*/
path1.moveTo(150, 200);//设置起点坐标
path1.lineTo(100, 300);//从前一个点到当前点的直线
path1.lineTo(200, 300);
path1.close();//使这些点构成封闭的多边形
```

Paint类常用的方法如下:
- setColor(int color),设置画笔颜色,参数为颜色值,可以为颜色常量,如Color.RED,也可以是16进制的颜色值,如0xF8F8FF00,对应0x|Red|Green|Blue|Alpha,每一个单位的取值范围[00,ff]。
- setStyle(Paint.Style style),设置画笔样式,有三个可选常量:Paint.Style.FILL(填充样式)、Paint.Style.STROKE(外边框线条样式)、Paint.Style.FILL_AND_STROKE(填充并带外边框线条样式)。
- setTextSize(float textSize),设置字体大小。

【例8.1】通过图像绘制的方法自定义一个组件,并进行事件响应测试。

(1)新建一个项目,项目名为DrawFigureDemo,在项目中新建一个Java Class,名称为CustomView,编写代码如下:

```
package cc.turbosnail.drawfiguredemo;
import android.content.Context;
import android.graphics.Canvas;
import android.graphics.Color;
import android.graphics.Paint;
import android.graphics.Path;
import android.util.AttributeSet;
import android.util.Log;
import android.view.View;
```

```java
public class CustomView extends View {
    private Paint paint;
    /*在Java代码中new的时候调用*/
    public CustomView(Context context) {
        super(context);
        paint = new Paint();
    }
    /*在XML布局文件中放置的时候调用*/
    public CustomView(Context context, AttributeSet attrs) {
        super(context, attrs);
        paint = new Paint();
    }
    @Override
    protected void onDraw(Canvas canvas) {
        super.onDraw(canvas);
        int width = canvas.getWidth();
        int height = canvas.getHeight();
        paint.setColor(Color.GRAY);
        canvas.drawRect(0, 0, width, height, paint);//绘制背景
        paint.setColor(Color.BLUE);
        Path path = new Path();
        /*设置多边形的点*/
        path.moveTo(0, height/2);//设置起点坐标
        path.lineTo(width/3, 0);//从前一个点到当前点的直线
        path.lineTo(width/3*2, 0);
        path.lineTo(width, height/2);
        path.lineTo(width/3*2, height);
        path.lineTo(width/3, height);
        path.close();//使这些点构成封闭的多边形
        canvas.drawPath(path, paint);//绘制多边形
    }
}
```

（2）在activity_main.xml文件中通过完整的名称使用自定义的View组件，即"包名.类名"的形式，示例如下：

```xml
<?xml version="1.0" encoding="utf-8"?>
<android.support.constraint.ConstraintLayout xmlns:android="http://schemas.android.com/apk/res/android"
```

```xml
    xmlns:app="http://schemas.android.com/apk/res-auto"
    xmlns:tools="http://schemas.android.com/tools"
    android:layout_width="match_parent"
    android:layout_height="match_parent"
    tools:context=".MainActivity">
    <cc.turbosnail.drawfiguredemo.CustomView
        android:id="@+id/c_view"
        android:layout_width="150dp"
        android:layout_height="100dp"
        app:layout_constraintBottom_toBottomOf="parent"
        app:layout_constraintLeft_toLeftOf="parent"
        app:layout_constraintRight_toRightOf="parent"
        app:layout_constraintTop_toTopOf="parent"/>
</android.support.constraint.ConstraintLayout>
```

（3）在MainActivity中获取CustomView对象，并实现简单的事件监听来测试这个自定义组件。

```java
package cc.turbosnail.drawfiguredemo;
import android.support.v7.app.AppCompatActivity;
import android.os.Bundle;
import android.view.View;
import android.widget.Toast;
public class MainActivity extends AppCompatActivity {
    CustomView customView;
    @Override
    protected void onCreate(Bundle savedInstanceState) {
        super.onCreate(savedInstanceState);
        setContentView(R.layout.activity_main);
        customView = findViewById(R.id.c_view);
        customView.setOnClickListener(new View.OnClickListener() {
            @Override
            public void onClick(View v) {
                Toast.makeText(MainActivity.this,"您点击了自定义组件",Toast.LENGTH_SHORT).show();
            }
        });
    }
}
```

(4)运行程序,可以看到自定义的组件如图 8.1 所示。

图 8.1 自定义组件

自定义组件是一个实用的功能,开发中可以通过继承的方式自定义 View、Button、TextView、布局等各种 UI 元素,以实现个性化的显示外观,提供更丰富的界面效果,如圆形头像图片、自定义动画效果等。

### 8.1.2 绘制图像

图像可以由 Bitamp 对象来表示,一张图片可以从文件系统、R.drawable 中的资源文件、数据流、字节数组等位置读入程序中,构建一个 Bitamp 对象,然后就可以进行图像绘制、图像处理(旋转、压缩、剪裁)、图像输出保存等操作。

(1)通过 BitmapFactory 构造 Bitamp 对象。

☞ 加载本地图片:

Bitmap bitmap = BitmapFactory.decodeFile(pathName);//pathName 为文件路径

☞ 加载项目资源文件中的图片:

Bitmap bitmap = BitmapFactory.decodeResource(getResources(), R.drawable.picName);//picName 为图片在资源文件中的名称

☞ 加载输入流图片:

Bitmap bitmap = BitmapFactory.decodeStream(inputStream);//inputStream 为图片数据的 InputStream 流,一般是异步从网络获取的

☞ 加载字节数组图片:

Bitmap bitmap = BitmapFactory.decodeByteArray(byteArray,0, byteArray.length);
//byteArray 为图片数据构成的字节数组

(2)绘制 Bitamp 对象。Canvas 提供了绘制 Bitamp 对象的方法:

drawBitmap(Bitmap bitmap, float left, float top, Paint paint)

例如,在 Ondraw 方法中绘制 mipmap 目录中的 Android 图标:

```
Bitmap bitmap = BitmapFactory.decodeResource(getResources(), R.mipmap.ic_launcher);
canvas.drawBitmap(bitmap,100,300,paint);
```

## 8.2 动画

动画在 Android 开发中比较常用,应用软件在加载、运行过程中加入适量动画效果,能带来更好的交互效果,提供更好的用户体验。

### 8.2.1 帧动画(Frame Animation)

帧动画的实现原理是:把图片资源顺次排列,每张图片代表一帧,然后依次播放显示出来,实现人眼看到的动画效果。帧动画通过<animation-list>标签定义,然后在 Android 代码中把它设置给显示组件作为背景或者数据源,如 ImageView 的背景等,最后由 AnimationDrawable 提供的 start()等方法进行播放控制。

[例8.2] 自定义并实现一个帧动画。

(1)新建一个项目,项目名为 FrameAnimDemo。把准备好的动画图片 ic_t1.png、ic_t2.png、ic_t2.png 复制到项目的 drawable 目录中。

(2)新建 XML 动画文件。在项目目录"res"→"drawable"文件夹上点击右键,然后选择"New"→"Drawable Resource File",填入文件名 frameani,如图 8.2 所示。

图 8.2 创建帧动画文件

修改文件,将根元素修改为<animation-list>标签,代码如下:

```xml
<?xml version="1.0" encoding="utf-8"?>
<animation-list xmlns:android="http://schemas.android.com/apk/res/android"
    android:oneshot="false">
    <item
        android:drawable="@drawable/ic_t1"
        android:duration="200"/>
    <item
        android:drawable="@drawable/ic_t2"
        android:duration="200"/>
    <item
        android:drawable="@drawable/ic_t3"
        android:duration="200"/>
</animation-list>
```

<animation-list>是根节点，android:oneshot 为 true 表示只播放一次，为 false 则表示循环播放，每一个<item>表示一帧，图片资源由 android:drawable 属性导入，android:duration 表示当前帧的显示时间，单位为毫秒。

（3）在 activity_main.xml 布局文件中放置一个图片视图（ImageView），id 命名为 iv_anim。将它的背景设置为动画文件 frameani，写法为 android:background="@drawable/frameani"。

完整代码清单如下：

```xml
<?xml version="1.0" encoding="utf-8"?>
<android.support.constraint.ConstraintLayout xmlns:android="http://schemas.android.com/apk/res/android"
    xmlns:app="http://schemas.android.com/apk/res-auto"
    xmlns:tools="http://schemas.android.com/tools"
    android:layout_width="match_parent"
    android:layout_height="match_parent"
    tools:context=".MainActivity">
    <ImageView
        android:id="@+id/iv_anim"
        android:layout_width="wrap_content"
        android:layout_height="wrap_content"
        android:background="@drawable/frameani"
        app:layout_constraintBottom_toBottomOf="parent"
        app:layout_constraintLeft_toLeftOf="parent"
        app:layout_constraintRight_toRightOf="parent"
        app:layout_constraintTop_toTopOf="parent"/>
</android.support.constraint.ConstraintLayout>
```

（4）在 MainActivity 中播放动画。通过 getBackground 方法获取图片视图的背景图，然后转换成 AnimationDrawable 对象，由该对象提供的方法进行动画播放相关操作。例如，在 onCreate 方法中直接播放动画的代码清单如下：

```java
ImageView imageView = findViewById(R.id.iv_anim);
AnimationDrawable frameAni = (AnimationDrawable)imageView.getBackground();
frameAni.start();
```

AnimationDrawable 还提供了其他一些进行动画播放控制的方法：

☞ public void stop()，停止播放动画。

☞ public boolean isRunning()，判断当前动画是否处于播放状态。

☞ public int getNumberOfFrames()，获取总的帧数。

☞ public Drawable getFrame(int index)，获取 index 索引位置的帧图片对象。

### 8.2.2 补间动画(Tween Animation)

补间动画的实现原理是:开发者指定动画开始、结束等关键帧,中间的过程将会依照关键帧通过计算自动补齐,以实现完整的动画效果。补齐方法包括淡入淡出(alpha)、位移(translate)、缩放(scale)、旋转(rotate)等。

补间动画作用于 View 上,在实现过程中组件的属性没有发生改变,即如果移动了一个 Button,但是要响应 Button 的事件还是需要在 Button 的初始位置去点击。

(1) 淡入淡出动画。淡入淡出动画是指控制显示元素的透明度发生渐变,产生动画效果。

**[例8.3]** 自定义并实现一个淡入淡出动画。

① 新建一个项目,项目名为 TweenAnimDemo。把需要添加动画效果的图片 ic_anim.png 复制到项目的 drawable 目录中。

② 创建放置动画资源的 anim 文件夹。在项目的 res 目录上点击右键,选择"New"→"Android resource directory",在弹出框中"Resource type"下拉列表中选择"anim"。

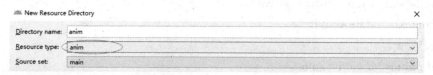

图 8.3 创建 anim 文件夹

③ 创建动画文件。在 anim 目录上点击右键,新建"Animation resource file",文件名为 anim_alpha,修改文件代码,实现<alpha>标签如下:

```
<?xml version="1.0" encoding="utf-8"?>
<alpha xmlns:android="http://schemas.android.com/apk/res/android"
    android:duration="3000"
    android:fromAlpha="1.0"
    android:toAlpha="0.0"
    android:repeatCount="-1"
    android:repeatMode="reverse"/>
```

其中,android:duration 为动画播放时间的毫秒数,android:fromAlpha、android:toAlpha 为透明度的变化值,取值范围为[0.0,1.0],android:repeatCount 表示重复次数,负数表示无限循环,android:repeatMode 表示重复模式,可取值为"reverse"和"restart"。

④ 构建 Animation 对象,为界面元素实现动画效果。

AnimationUtils 工具类提供了类方法 loadAnimation()来创建动画对象,该方法有两个参数,一个是当前上下文环境 Contex,一个是动画文件的 id。View 组件可以通过 startAnimation 方法来启动动画,既可以是组件对象,也可以是布局对象。

```
//为UI组件实现动画
Animation animation = AnimationUtils.loadAnimation(this, R.anim.anim_alpha);
```

ImageView img = findViewById(R.id.iv_anim);
img.startAnimation(animation);
//为布局实现动画
ConstraintLayout layout =findViewById(R.id.ly_main);
Animation animation = AnimationUtils.loadAnimation(this, R.anim.anim_alpha);
layout.startAnimation(animation);

⑤ 运行程序，可以看到图片出现了淡入淡出的动画效果，如图8.4所示。

图8.4 淡入淡出动画运行效果

（2）位移动画。位移动画是指控制元素的位置发生渐变，产生动画效果。一些动画效果设置的属性如表8.1所示。

表8.1 位移动画的属性设置列表

| 属性 | 描述 | 取值 |
| --- | --- | --- |
| android:duration | 动画持续时间 | 毫秒数，值为整数型，如"2000"可选三种值： |
| android:fromXDelta | 起点的X坐标 | （1）数值型：如"100"，单位是默认的px； |
| android:fromYDelta | 起点的Y坐标 | （2）百分比：如"50%"，是相对于自身宽度的百分比值； |
| android:toXDelta | 终点的X坐标 | （3）百分比+"p"：如"50%p"，是相对于自己父容器组件的百分比值 |
| android:toYDelta | 终点的Y坐标 | |
| android:fillAfter | 动画结束后是否保持最终状态 | "true"表示保持，"false"表示不保持 |
| android:fillBefore | 动画结束后是否恢复到开始前的状态 | "true"表示恢复开始前的状态，"false"表示不恢复 |
| android:repeatCount | 动画重复的次数 | 整数型，如"10"。负数表示无限循环，如"-1" |
| android:repeatMode | 重复模式，有两种 | "reverse"，倒序重播；"restart"，顺序重播 |

位移动画的实现过程与淡入淡出动画类似，首先在项目的anim目录中创建XML动画文件"Animation resource file"，如文件命名为anim_translate，然后实现动画<translate>标签，示例代码如下：

```xml
<?xml version="1.0" encoding="utf-8"?>
<translate xmlns:android="http://schemas.android.com/apk/res/android"
android:interpolator="@android:anim/accelerate_decelerate_interpolator"
android:fromXDelta="0"
android:toXDelta="100"
android:fromYDelta="0"
android:toYDelta="100"
android:duration="2000"
android:repeatCount="-1"
android:repeatMode="reverse"/>
```

最后加载并为界面元素实现动画效果,其实现方法与淡入淡出动画相同,示例代码如下:

```
Animation animation = AnimationUtils.loadAnimation(this, R.anim.anim_translate);
ImageView img = findViewById(R.id.iv_anim);
img.startAnimation(animation);
```

(3)缩放动画。缩放动画是指控制元素的大小发生渐变,产生动画效果。一些动画效果设置的属性如表8.2所示。

表8.2 缩放动画的属性设置列表

| 属性 | 描述 | 取值 |
| --- | --- | --- |
| android:fromXScale | 起始时X轴方向动画控件的大小 | 可选三种值:<br>(1)数值型:如"0.3"表示0.3倍;<br>(2)百分比:如"50%",是相对于自身宽度的百分比值;<br>(3)百分比+"p":如"50%p",是相对于自己父容器组件的百分比值 |
| android:fromYScale | 起始时Y轴方向动画控件的大小 | |
| android:toXScale | 结束时X轴方向动画控件的大小 | |
| android:toYScale | 结束时Y轴方向动画控件的大小 | |
| android:pivotX | 缩放中心坐标的X值 | |
| android:pivotY | 缩放中心坐标的Y值 | |

首先在项目目录的anim文件夹中创建缩放动画的XML文件,命名为anim_Scale,然后实现<scale>标签,示例代码如下:

```xml
<?xml version="1.0" encoding="utf-8"?>
<scale xmlns:android="http://schemas.android.com/apk/res/android"
    android:fromXScale="0.0"
    android:toXScale="1.0"
    android:fromYScale="0.0"
    android:toYScale="1.0"
    android:pivotX="50%"
    android:pivotY="50%"
    android:duration="2000"/>
```

(4)旋转动画。旋转动画是指控制显示元素的旋转状态发生渐变,产生动画效果。一些动画效果设置的属性如表 8.3 所示。

表 8.3　旋转动画的属性设置列表

| 属性 | 描述 | 取值 |
| --- | --- | --- |
| Android:fromDegrees | 旋转开始的起始角度 | 度数值,"0"~"360"之间 |
| Android:toDegrees | 旋转结束的终止角度 | |
| Android:pivotX | 旋转中心的 X 坐标 | 同上,pivotX、pivotY |
| Android:pivotY | 旋转中心的 Y 坐标 | |
| android:visible | 初始时是否显示 | "true",显示;"false",不显示 |

首先在项目目录的 anim 文件夹中创建旋转动画的 XML 文件,命名为 anim_rotate,然后实现<rotate>标签,示例代码如下:

```
<?xml version="1.0" encoding="utf-8"?>
<rotate xmlns:android="http://schemas.android.com/apk/res/android"
    android:fromDegrees="0"
    android:toDegrees="360"
    android:pivotX="50%"
    android:pivotY="50%"
    android:visible = "true"
    android:duration = "2000"
    android:repeatCount = "-1"/>
```

(5)混合使用。以上四种补间动画可以组合使用,以实现灵活多样的动画效果。在组合使用时,创建的"Animation resource file"动画文件根目录保持<set>标签,然后在其中添加其他动画标签的组合并实现即可,其代码框架如下:

```
<?xml version="1.0" encoding="utf-8"?>
<set xmlns:android="http://schemas.android.com/apk/res/android">
    <scale
       ……/>
    <rotate
       …… />
    <translate
       …… />
    <alpha
       ……/>
</set>
```

另外,在补间动画中,android:interpolator 属性可以提供插值速率控制的效果,Android

提供了很多运动效果，例如：

- "@android:anim/accelerate_interpolator"，加速运动（播放越来越快）。
- "@android:anim/decelerate_interpolator"，减速运动（播放越来越慢）。
- "@android:anim/accelerate_decelerate_interpolator"，先加速再减速。
- "@android:anim/anticipate_interpolator"，弹簧效果，先反向，再加速返回。
- "@android:anim/anticipate_overshoot_interpolator"，同上弹簧效果，结束有回弹。
- "@android:anim/bounce_interpolator"，弹球效果，结束后回弹几下。
- "@android:anim/cycle_interpolator"，按次数循环播放，速率沿正弦曲线改变。
- "@android:anim/linear_interpolator"，线性以常量速率改变。
- "@android:anim/overshoot_interpolator"，加速播放，结束后回弹。

**[项目案例22]** 在AccountBook项目为欢迎页的logo图像添加动画效果

程序启动时，logo图像从小变大旋转出现，并以淡入淡出的效果进行显示。

（1）在项目的res目录上点击右键，选择"New"→"Android resource directory"，在弹出框"Resource type"下拉列表中选择"anim"，如图8.5所示。

图8.5 创建AccountBook项目的anim文件夹

（2）创建动画文件。在anim目录上点击右键，新建"Animation resource file"，文件名为anim_set，在<set>标签中添加动画组合如下：

```xml
<?xml version="1.0" encoding="utf-8"?>
<set xmlns:android="http://schemas.android.com/apk/res/android">
    <scale
        android:duration="2000"
        android:fromXScale="0.0"
        android:fromYScale="0.0"
        android:pivotX="50%"
        android:pivotY="50%"
        android:toXScale="1.0"
        android:toYScale="1.0"/>
    <rotate
        android:duration="1500"
        android:fromDegrees="0"
        android:pivotX="50%"
        android:pivotY="50%"
```

```
        android:repeatCount="-1"
        android:toDegrees="360"
        android:visible="true"/>
    <alpha
        android:duration="3000"
        android:fromAlpha="0.0"
        android:repeatCount="-1"
        android:repeatMode="reverse"
        android:toAlpha="1.0"/>
</set>
```

（3）在 WelcomeActivity 的 onCreate 方法中添加动画实现的代码如下：

```
Animation animation = AnimationUtils.loadAnimation(this, R.anim.anim_set);
ImageView img = findViewById(R.id.iv_logo);
img.startAnimation(animation);
```

其中，iv_logo 为 activity_welcome.xml 布局文件中图片视图的 id。运行程序，可以看到 logo 图片在淡入淡出效果中一边旋转一边从小到大变化显示出来。

### 8.2.3 属性动画

属性动画是指动画元素的某些属性真实发生变化，渐变过程中产生的动画效果。属性动画可以实现比补间动画更多样的自定义动画效果，同时不会出现补间动画中属性其实没有真实改变的情况，而且它可以针对组件的某个具体属性甚至非组件的对象定义动画，是制作复杂、细致动画效果时较常用的一种动画实现方式。

属性动画的实现原理：由数值发生器（ValueAnimator）设置开始和结束的两个数值，通过它的 setDuration( ) 方法设置动画总时间，通过插值器（Intepolator）来控制动画速率，数值发生器会根据这些设置自动生成一个数值序列，然后由它的 getAnimatedValue( ) 方法获取数值序列值，动画效果就可以通过这个序列值不断改变对象的某些属性值来实现。示例代码如下：

```
final ValueAnimator animator = ValueAnimator.ofFloat(1, 0);
animator.setDuration(1000);
animator.setInterpolator(new AccelerateInterpolator());
animator.addUpdateListener(new ValueAnimator.AnimatorUpdateListener() {
    @Override
    public void onAnimationUpdate(ValueAnimator valueAnimator) {
        float value = (float)valueAnimator.getAnimatedValue();
        imageView.setAlpha(value);
    }
});
animator.start();
```

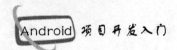

为了简化属性动画的实现过程，Android 提供了 ObjectAnimator 类，它封装了 ValueAnimator 接口实现，并简化了属性动画的实现代码。其实现过程如下：

```
ObjectAnimator animator = ObjectAnimator.ofFloat(imageView,"alpha",1,0);
animator.setDuration(1000);
animator.setInterpolator(new AccelerateIntepolator());
animator.start();
```

其中，ofFloat()的方法声明如下：

public static ObjectAnimator ofFloat（Object target，String propertyName，float…values）；

当需要为某个属性设置属性动画时，ObjectAnimator.ofFloat()方法的第二个参数 propertyName 就是该属性的访问器方法 setXXX 去掉"set"后首字母小写的字符串，如"alpha"来自于 imageView 对象自身属性设置的方法 setAlpha()。View 常见属性和 ObjectAnimator.ofFloat()第二个参数的对应关系如下：

- public void setAlpha(float alpha)：alpha
- public void setRotation(float rotation)：rotation
- public void setRotationX(float rotationX)：rotationX
- public void setRotationY(float rotationY)：rotationY
- public void setTranslationX(float translationX)：translationX
- public void setTranslationY(float translationY)：translationY
- public void setScaleX(float scaleX)：scaleX
- public void setScaleY(float scaleY)：scaleY

setInterpolator()方法设置插值器（Intepolator），插值器负责控制动画变化的速率，使动画效果能够以匀速、加速、减速、抛物线速率等各种速率进行变化，常见的插值器如下：

- LinearInterpolator，以常量速率线性改变。
- AccelerateInterpolator，加速运动（播放越来越快）。
- DecelerateInterpolator，减速运动（播放越来越慢）。
- AccelerateDecelerateInterpolator，先加速再减速。
- CycleInterpolator，动画循环播放特定的次数，速率沿着正弦曲线改变。
- BounceInterpolator，弹球效果，结束后回弹几下。
- OvershootInterpolator，加速播放，结束后回弹。
- AnticipateInterpolator，弹簧效果，先反向，再加速返回。
- AnticipateOvershootInterpolator，同上弹簧效果，结束有回弹。

## 8.3 跳转动画

Android 提供了界面跳转动画，在 Android 5.0 以上的版本中，可以使得 Activity 切换时有类似于幻灯片切换一样的效果，四种跳转动画如下：

① Explode，界面元素从各个方向进入。
② Slide，界面元素从底部依次向上运动。
③ Fade，界面元素渐变出现。
④ Share，前后界面设置成共享的元素通过渐变切换。

界面跳转动画的实现过程如下：

(1) 启动跳转时声明使用跳转动画，代码如下：

```
Intent intent = new Intent(MainActivity.this,SecondActivity.class);
startActivity(intent,ActivityOptions.makeSceneTransitionAnimation(MainActivity.this).toBundle());
```

(2) 在目标界面SecondActivity中添加动画效果，代码如下：

```
protected void onCreate(Bundle savedInstanceState) {
    super.onCreate(savedInstanceState);
    getWindow().setEnterTransition(new Explode());
    setContentView(R.layout.activity_second);
}
```

类似地，可以添加其他两种动画效果，代码如下：

```
//添加Slide效果
getWindow().setEnterTransition(new Slide());
//添加Fade效果
getWindow().setEnterTransition(new Fade());
getWindow().setExitTransition(new Fade());
```

Share效果因为要设置"共享元素"，所以写法有所不同，它的实现过程如下：

(1) 为需要共享效果的元素设置相同的android:transitionName属性。如对两个Activity布局中的元素设置相同的transitionName属性。

OneActivity的布局文件activity_one.xml中的按钮，代码如下：

```
<Button android:id="@+id/btn_1"
    android:layout_width="100dp"
    android:layout_height="100dp"
    android:transitionName="sharedView1"/>
```

TwoActivity的布局文件activity_two.xml中的按钮，代码如下：

```
<Button android:id="@+id/btn_2"
    android:layout_width="100dp"
    android:layout_height="100dp"
    android:transitionName="sharedView1"/>
```

（2）跳转声明。其中，makeSceneTransitionAnimation方法的第二个参数是第一个Activity中的View对象id，第三个参数是android:transitionName的属性值，代码如下：

```
Button btnOne = findViewById(R.id.btn_1);
Intent intent = new Intent(OneActivity.this,TwoActivity.class);
startActivity(intent, ActivityOptions. makeSceneTransitionAnimation(OneActivity. this, btnOne, "sharedView1").toBundle());
```

如果有多个共享的组件，则可以用Pair封装，代码示例如下：

```
startActivity(intent, ActivityOptions.makeSceneTransitionAnimation(OneActivity.this, Pair.create (btn1, "sharedView1"), Pair.create(view, "sharedView2"), Pair.create(tv, "sharedView3")) .toBundle());
```

（3）在目标Activity中开启Transition模式。例如在TwoActivity中添加如下代码：

```
protected void onCreate(Bundle savedInstanceState) {
    super.onCreate(savedInstanceState);
    getWindow().requestFeature(Window.FEATURE_CONTENT_TRANSITIONS);
    setContentView(R.layout.activity_two);
}
```

## [项目案例23] 在AccountBook项目为账目列表主界面打开时添加跳转动画效果

当登录成功后跳转到账目列表主界面时，主界面上的元素以从各个方向进入的效果出现。

（1）打开AccountBook项目的LoginActivity代码，找到登录验证成功后Activity跳转的代码段：

```
Intent intent = new Intent(LoginActivity.this, MainActivity.class);
startActivity(intent);
LoginActivity.this.finish();
```

将其修改为：

```
Intent intent = new Intent(LoginActivity.this, MainActivity.class);
//检查系统是否为Android 5.0以上
if (Build.VERSION.SDK_INT >= Build.VERSION_CODES.LOLLIPOP) {
    startActivity(intent, ActivityOptions. makeSceneTransitionAnimation(LoginActivity. this). toBundle());
} else {
    startActivity(intent);
}
```

注意，需要对Android系统是否为5.0以上进行判断，Activity的切换效果只针对5.0以上版本有效。

（2）打开跳转目标MainActivity的代码，在setContentView方法之前添加动画效果的实现，这里也需要对Android系统的版本进行判断，代码如下：

```
//检查系统是否为Android 5.0以上
if (Build.VERSION.SDK_INT >= Build.VERSION_CODES.LOLLIPOP) {
    getWindow().setEnterTransition(new Explode().setDuration(2000));
}
setContentView(R.layout.activity_main);
```

（3）运行程序，可以看到，在登录成功后跳转到列表主界面时，主界面上的列表、按钮等元素从屏幕的四个方向飞入进来，构成完整的界面。

## 8.4 基于SurfaceView的动画

通过在View中绘制图像，并不断修改画面（图形图像的位置、大小、增减、修改等）再重绘View，这样就可实现自定义的动画效果，即按照一定的频率不断修改画面上的某些属性，然后通过View让onDraw()方法不断地执行，界面被反复重绘来实现动画，其中，onDraw()方法的调用需要通过invalidate()或者postinvalidate()方法来实现。这样的自定义动画在动画制作、游戏动画开发中大量使用。但是，由于View没有双缓存机制，为了防止出现一屏尚未绘制完成、另一屏已经开始绘制而导致的画面撕裂问题，开发者必须自行实现一个缓存View，等到全部绘制完成后再赋给界面显示。同时，如果参数修改的操作放在子线程中，子线程直接要更新UI主线程也是不允许的，这就需要频繁地进行子线程和UI主线程通信，于是增加了动画实现的工作量。为了适应画面不断刷新的绘图需要，Android专门提供了SurfaceView来更便捷地实现动画和游戏绘图。

SurfaceView类是View的子类，它内嵌了一个用于绘图的Surface，且自身实现了双缓存、在子线程中更新画面等功能，具有较高的绘图效率，其绘图原理是：每次绘图前先通过lockCanvas()锁定画板，这个画板是在缓存区，并不在真实的显示界面上，绘图结束后，通过unlockCanvasAndPost()方法将缓存区画好的一屏全部输出到真实的显示界面上，而原先界面上的一屏将被传回给缓存区。

SurfaceView绘图的过程由SurfaceHolder接口来进行监听，SurfaceHolder提供了lockCanvas()、unlockCanvasAndPost()等方法，除此之外，它还提供了addCallback(SurfaceHolder.Callback callback)方法，用来添加一个SurfaceHolder.Callback接口，SurfaceHolder.Callback提供了监听绘图区Surface状态变化时的回调方法如下：

- public void sufaceChanged(SurfaceHolder holder, int format, int width, int height){};//Surface的大小发生改变时回调
- public void surfaceCreated(SurfaceHolder holder){};//Surface创建时回调，绘图线程一般在这里启动

☞ public void surfaceDestroyed(SurfaceHolder holder){};//Surface销毁时回调,一般在这里停止线程、释放资源

[例8.4] 自定义并实现一个基于SurfaceView的动画。在SurfaceView中绘制一个图像,并让它从屏幕右边向左边平行移动。

(1)新建一个项目,项目名为SurfaceViewAnimDemo。把需要添加动画效果的图片turbosnail.png复制到项目中的drawable目录中。

(2)在项目目录"java"→"cc.turbosnail.surfaceviewanimdemo"中点击右键,新建一个"Java Class",命名为SurfaceViewAnim,并实现代码如下:

```
package cc.turbosnail.surfaceviewanimdemo;
import android.content.Context;
import android.graphics.Bitmap;
import android.graphics.BitmapFactory;
import android.graphics.Canvas;
import android.graphics.Color;
import android.graphics.Paint;
import android.view.SurfaceHolder;
import android.view.SurfaceView;
import android.view.WindowManager;
public class SurfaceViewAnim extends SurfaceView implements SurfaceHolder.Callback,Runnable{
    private SurfaceHolder mHolder;//声明SurfaceHolder
    private Canvas mCanvas;
    private boolean isDrawing;//控制绘图线程
    private Paint paint;//画笔对象
    private Bitmap snail;//实现动画的图片对象
    private int snail_X;//动画元素的起始X坐标
    public SurfaceViewAnim(Context context) {
        super(context);
        snail = BitmapFactory.decodeResource(getResources(), R.drawable.turbosnail);
        WindowManager wm = (WindowManager) context.getSystemService(Context.WINDOW_SERVICE);
        int width = wm.getDefaultDisplay().getWidth();
        snail_X=width;//起始X坐标赋值为当前屏幕宽度值,即屏幕最右边
        paint=new Paint();
        mHolder = this.getHolder();//获取SurfaceHolder对象
        mHolder.addCallback(this);//添加Callback接口
    }
    @Override
```

```java
public void surfaceCreated(SurfaceHolder holder) {
    isDrawing = true;
    new Thread(this).start();//启动绘图的线程
}
@Override
public void surfaceChanged(SurfaceHolder holder, int format, int width, int height) {
}
@Override
public void surfaceDestroyed(SurfaceHolder holder) {
    isDrawing = false;//终止绘图线程
}
public void run() {
    while(isDrawing) {
        snail_X--;//绘图元素属性修改,此处为修改其X坐标
        if(snail_X<=0){//到达最左边后,不再移动
            snail_X=0;
        }
        mydraw();//绘图
        try {
            Thread.sleep(30);
        } catch(Exception e) {
        }
    }
}
private void mydraw() {
    try {
        mCanvas = mHolder.lockCanvas();//锁定画板,开始绘图
        paint.setColor(Color.WHITE);
        mCanvas.drawRect(0,0,getWidth(),getHeight(),paint);//绘制背景
        mCanvas.drawBitmap(snail,snail_X,100,paint);//绘图
    } catch (Exception e) {
    } finally {
        if (mCanvas != null){
            mHolder.unlockCanvasAndPost(mCanvas);//解锁画板,推送到界面
        }
    }
}
}
```

（3.）在 MainActivity 中创建 SurfaceViewAnim 对象，并把这个对象设置为当前显示视图，代码如下：

```
package cc.turbosnail.surfaceviewanimdemo;
import android.support.v7.app.AppCompatActivity;
import android.os.Bundle;
public class MainActivity extends AppCompatActivity {
    SurfaceViewAnim anim;
    @Override
    protected void onCreate(Bundle savedInstanceState) {
        super.onCreate(savedInstanceState);
        anim=new SurfaceViewAnim(this);
        setContentView(anim);//设置 anim 为显示视图
    }
}
```

（4）运行程序，如图 8.6 所示，可以看到，图片在屏幕上从左往右水平移动，当到达最左边时停止移动。

图 8.6　SurfaceView 动画

本例在线程中不断地把一个图片对象的 X 坐标进行修改，实现了图片水平移动的动画效果，除此之外，还可以添加多个图片或者其他可绘制的图形图像（如圆形、组合的图形等）作为动画元素，它们的各种可变属性都可以定义成动画变量，在线程中修改变量值，然后再绘制以实现丰富多变的效果。也可以通过设置状态变量，当达到某一个条件时，再绘制某些动画元素，这样就可以实现具有先后顺序的连贯动画，实现动画片的制作。

# 第9章 手机传感器概述

传感器可以探测、感知外界信号,并将得到的信息传递出来,开发者可以根据这些信息实现各种基于传感器的应用开发。传感器在生产、生活中被广泛地使用,当前的手机一般都会内置多种传感器,例如加速度传感器、陀螺仪传感器、磁场传感器、方向传感器、光线传感器、温度传感器、湿度传感器、压力传感器、重力传感器、接近传感器、线性加速度传感器等。Android系统提供了对传感器的支持,能够获取传感器硬件的信号变化,并提供了开发者能调用的传感器开发接口。基于Android传感器应用,我们可以开发出丰富多彩、个性化的应用程序。

## 9.1 传感器的使用方法

(1) 创建传感器服务类(SensorManager)对象。
SensorManager sensorManage=(SensorManager)getSystemService(SENSOR_SERVICE);//getSystemService为Context提供的方法,根据参数SENSOR_SERVICE得到传感器服务类的对象
(2) 创建指定类型的传感器对象。调用SensorManager的getDefaultSensor(int type)方法创建传感器对象,参数type为传感器的类型,常见的传感器参数如表9.1所示。代码如下:
Sensor sensor = sensorManager.getDefaultSensor(Sensor.TYPE_LIGHT);

表9.1 传感器常量列表

| 序号 | 传感器 | TYPE 常量 |
| --- | --- | --- |
| 1 | 加速度传感器 | Sensor.TYPE_ACCELEROMETER |
| 2 | 温度传感器 | Sensor.TYPE_AMBIENT_TEMPERATURE |
| 3 | 陀螺仪传感器 | Sensor.TYPE_GYROSCOPE |
| 4 | 光线传感器 | Sensor.TYPE_LIGHT |
| 5 | 磁场传感器 | Sensor.TYPE_MAGNETIC_FIELD |
| 6 | 压力传感器 | Sensor.TYPE_PRESSURE |
| 7 | 临近传感器 | Sensor.TYPE_PROXIMITY |
| 8 | 湿度传感器 | Sensor.TYPE_RELATIVE_HUMIDITY |
| 9 | 重力传感器 | Sensor.TYPE_GRAVITY |
| 10 | 线性加速传感器 | Sensor.TYPE_LINEAR_ACCELERATION |
| 11 | 旋转向量传感器 | Sensor.TYPE_ROTATION_VECTOR |

（3）为传感器对象监听注册。调用SensorManager的registerListener()方法进行监听注册，一般情况下建议放在Activity的OnResume()方法中。该方法声明如下：

public boolean registerListener(SensorEventListener listener, Sensor sensor, int samplingPeriodUs)，三个参数分别为：

- listener：为传感器注册的监视器对象，它实现SensorEventListener接口，在接口方法中对传感器事件进行捕获和处理，它的两个接口方法如下：

    public void onSensorChanged(SensorEvent event) { }，获取到传感器事件时回调，event对象中封装了传感器返回的各项数据，通过float[] values = event.values;可以得到一个数据数组。

    public void onAccuracyChanged(Sensor sensor, int accuracy) { }，传感器精度发生变化时回调。

- sensor：被监听的传感器对象。

- samplingPeriodUs：传感器获取数据的频率，它包含如下几种频率常量：

    ✓ SensorManager.SENSOR_DELAY_FASTEST：最快，延迟最小，耗电量大，对数据处理速度要求也最高。

    ✓ SensorManager.SENSOR_DELAY_GAME：适合游戏频率，满足一般实时性要求的应用。

    ✓ SensorManager.SENSOR_DELAY_NORMAL：正常频率，满足实时性要求不高的应用。

    ✓ SensorManager.SENSOR_DELAY_UI：适合普通用户界面UI变化的频率，延迟较大，耗电量少，实时性要求比较低时应用。

（4）停止传感器事件监听。由于传感器事件频率高，而且activity变为不可见时依然在工作，所以应该及时停止，一般情况下建议放在Activity的onPause()方法中停止监听，方法声明如下：

- public void unregisterListener（SensorEventListener listener, Sensor sensor），取消对sensor的listener监听。

- public void unregisterListener（SensorEventListener listener），取消所有listener监听。

    例如如下代码：

```
public void onPause() {
    super.onPause();
    sensorManager.unregisterListener(this);//this 为实现了 SensorEventListener 接口的当前 Activity
}
```

## 9.2 常用传感器

传感器涉及坐标时,它的坐标系为:z轴指向天空,垂直于地面;y轴在设备的当前位置与地面相切,指向磁北极;x轴在设备的当前位置与地面相切,并且大致指向东。手机或者其他设备在水平放置时,它的坐标系为:x轴水平指向右,y轴垂直向上,z轴与屏幕垂直指向屏幕正面之外,如图9.1所示。

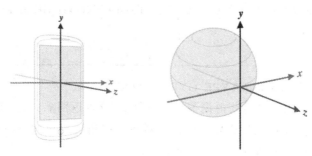

图9.1 传感器坐标系

(1) 加速度传感器(Sensor.TYPE_ACCELEROMETER)

加速度用来描述物体运动速度变化快慢,以 m/s² 为单位。在静止时加速度返回的值为地表静止物体的重力加速度,约为9.8 m/s²。由于重力的作用是向下的,所以当手机平放、竖直、横立时,z方向、y方向、x方向的重力加速度约为9.8 m/s²(正向)或者−9.8 m/s²(反向),当手机发生晃动时,对应方向的加速度也发生变化。所以,通过获取加速度值的变化,可以实现手机摇一摇等功能,一个简单的摇一摇功能实现如下:

```java
public void onSensorChanged(SensorEvent event) {
    float[] values = event.values;
    //获得x,y,z加速度
    float x = values[0];
    float y = values[1];
    float z = values[2];
    if(x>15 || y>15 || z>15){ //在开发中建议以常量的形式定义最小加速度
        Toast.makeText(this,"摇一摇",Toast.LENGTH_SHORT).show();
        //……//功能代码
    }
}
```

(2) 重力传感器(Sensor.TYPE_GRAVITY)

重力传感器可以得到(x,y,z)三个方向的重力值,在水平放置的状态下,它和静止时的加速度传感器返回值相同。手机倾斜时,对应坐标方向的重力会出现正负值的变化,如将手机向右倾斜,x会成为负值。利用这些特性,可以对手机的屏幕方向进行判断,例如:

```
public void onSensorChanged(SensorEvent event) {
    float[] values = event.values;
    //获得x,y,z方向重力值
    float x = values[0];
    float y = values[1];
    float z = values[2];
    if(z<0){
        Toast.makeText(this,"当前手机屏幕朝下",Toast.LENGTH_SHORT).show();
    }
}
```

（3）陀螺仪传感器(Sensor.TYPE_GYROSCOPE)

陀螺仪传感器可以精确地获得设备的旋转角度,分别返回$x$、$y$、$z$三个坐标轴的旋转角速度,单位是弧度/t。高精确度使得陀螺仪传感器在一些精度要求比较高的应用中广泛使用,如游戏开发中。陀螺仪传感器的简单示例如下:

```
public void onSensorChanged(SensorEvent event) {
    float[] values = event.values;
    float axisX = event.values[0];
    float axisY = event.values[1];
    float axisZ = event.values[2];
    //计算角速度
    float omegaMagnitude = sqrt(axisX*axisX + axisY*axisY + axisZ*axisZ);
}
```

（4）旋转向量传感器(Sensor.TYPE_ROTATION_VECTOR)

旋转向量传感器获取设备旋转角度和坐标轴的矢量值,获取到的数组值分别对应values[0]:$x\times\sin(\theta/2)$、values[1]:$y\times\sin(\theta/2)$、values[2]:$z\times\sin(\theta/2)$、values[3]:$\cos(\theta/2)$、values[4]:方向精度值。

（5）线性加速传感器(Sensor.TYPE_LINEAR_ACCELERATION)

线性加速传感器可以获取不包含重力的沿三个方向轴的加速度,三个值返回在( )中。

（6）湿度传感器(Sensor.TYPE_RELATIVE_HUMIDITY)

湿度传感器获取相对环境空气湿度百分比,存放在values[0]中,可以配合温度传感器计算出绝对湿度值和露点等。

（7）光线传感器(Sensor.TYPE_LIGHT)

光线传感器可以获取设备周围光线强度,单位是勒克斯(Lux)。该值返回在values[0]中。

（8）温度传感器(Sensor.TYPE_AMBIENT_TEMPERATURE)

温度传感器可以获取设备所处环境的温度,单位是摄氏度(°C)。该值返回在values[0]中。

（9）压力传感器(Sensor.TYPE_PRESSURE)

压力传感器可以获取设备得到压力的大小,单位是毫巴。该值返回在values[0]中。触屏是压力传感器的一个典型应用。

(10) 磁场传感器(Sensor.TYPE_MAGNETIC_FIELD)

磁场传感器可以获取坐标系三个方向的磁场值,单位是微特斯拉。$x$、$y$、$z$方向的磁场值分别返回在values[0]、values[1]、values[2]中。磁场传感器可用于开发指南针等,同时,它和加速度传感器配合使用作为SensorManager.getOrientation()方法的参数可以用来获取方向。示例代码如下:

```
float[] values = new float[3];//存放最终结果
float[] gravity = new float[3];//存放加速度传感器的值
float[] geomagnetic = new float[3];//存放磁场传感器的值
float[] r = new float[9];//getRotationMatrix参数,旋转矩阵
if(event.sensor.getType()==Sensor.TYPE_MAGNETIC_FIELD){
    geomagnetic=event.values;
}
if(event.sensor.getType()==Sensor.TYPE_ACCELEROMETER){
    gravity=event.values;
}
SensorManager.getRotationMatrix(r,null,gravity,geomagnetic);
SensorManager.getOrientation(r,values);
double azimuth=Math.toDegrees(values[0]);//z轴方向
double pitch=Math.toDegrees(values[1]);//x轴方向
double roll=Math.toDegrees(values[2]);//y轴方向
```

## 9.3 传感器使用示例与测试

[**例9.1**] 获取手机传感器的值,并输出显示到手机屏幕上。

(1) 新建一个项目,项目名为SensorDemo。

(2) 在布局文件main_activity.xml中放置3个文本标签,分别用来显示加速度、光线、磁场值。

```xml
<?xml version="1.0" encoding="utf-8"?>
<android.support.constraint.ConstraintLayout xmlns:android="http://schemas.android.com/apk/res/android"
    xmlns:app="http://schemas.android.com/apk/res-auto"
    xmlns:tools="http://schemas.android.com/tools"
    android:layout_width="match_parent"
    android:layout_height="match_parent"
```

```xml
        tools:context="cc.turbosnail.sensordemo.MainActivity">
    <TextView
        android:id="@+id/tv_accelerometer"
        android:layout_width="wrap_content"
        android:layout_height="wrap_content"
        android:text="正在获取加速度"
        app:layout_constraintBottom_toBottomOf="parent"
        app:layout_constraintLeft_toLeftOf="parent"
        app:layout_constraintRight_toRightOf="parent"
        app:layout_constraintTop_toTopOf="parent"
        android:layout_marginBottom="300dp"/>
    <TextView
        android:id="@+id/tv_light"
        android:layout_width="wrap_content"
        android:layout_height="wrap_content"
        android:text="正在获取光线值"
        app:layout_constraintTop_toBottomOf="@+id/tv_accelerometer"
        app:layout_constraintRight_toRightOf="parent"
        app:layout_constraintLeft_toLeftOf="parent"
        android:layout_marginTop="100dp"/>
    <TextView
        android:id="@+id/tv_magnetic"
        android:layout_width="wrap_content"
        android:layout_height="wrap_content"
        android:text="正在获取磁场值"
        app:layout_constraintTop_toBottomOf="@+id/tv_light"
        app:layout_constraintRight_toRightOf="parent"
        app:layout_constraintLeft_toLeftOf="parent"
        android:layout_marginTop="100dp"/>
</android.support.constraint.ConstraintLayout>
```

（3）在 MainActivity 中获取传感器的值，并把这些值显示到对应的文本标签上，代码如下：

```java
package cc.turbosnail.sensordemo;
import android.hardware.Sensor;
import android.hardware.SensorEvent;
import android.hardware.SensorEventListener;
import android.hardware.SensorManager;
```

```java
import android.support.v7.app.AppCompatActivity;
import android.os.Bundle;
import android.widget.TextView;
public class MainActivity extends AppCompatActivity implements SensorEventListener {
    private SensorManager sensorManager;
    private TextView tvAccelerometer, tvLight, tvMagnetic;
    @Override
    protected void onCreate(Bundle savedInstanceState) {
        super.onCreate(savedInstanceState);
        setContentView(R.layout.activity_main);
        initView();
        sensorManager = (SensorManager) getSystemService(SENSOR_SERVICE);//获取服务
    }
    public void initView(){
        tvAccelerometer = findViewById(R.id.tv_accelerometer);
        tvLight = findViewById(R.id.tv_light);
        tvMagnetic = findViewById(R.id.tv_magnetic);
    }
    @Override
    protected void onResume() {
        super.onResume();
        //分别为加速度、光线、磁场传感器注册监听器
        sensorManager.registerListener(this, sensorManager.getDefaultSensor
                (Sensor.TYPE_ACCELEROMETER),
                sensorManager.SENSOR_DELAY_GAME);
        sensorManager.registerListener(this, sensorManager.getDefaultSensor
                (Sensor.TYPE_LIGHT),
                sensorManager.SENSOR_DELAY_GAME);
        sensorManager.registerListener(this, sensorManager.getDefaultSensor
                (Sensor.TYPE_MAGNETIC_FIELD),
                sensorManager.SENSOR_DELAY_GAME);
    }
    @Override
    protected void onPause() {
        super.onPause();
        sensorManager.unregisterListener(this);//注销监听器
    }
    @Override
    public void onSensorChanged(SensorEvent event) {
```

```java
            int type = event.sensor.getType();
            float[] values = event.values;
            switch(type){
                case Sensor.TYPE_ACCELEROMETER:
                    tvAccelerometer.setText("加速度传感器(X,Y,Z):"+String.valueOf(values[0])+";"
                            +String.valueOf(values[1])+";"+String.valueOf(values[2]));
                    break;
                case Sensor.TYPE_LIGHT:
                    tvLight.setText("光线传感器:"+String.valueOf(values[0])+"Lux");
                    break;
                case Sensor.TYPE_MAGNETIC_FIELD:
                    tvMagnetic.setText("磁场传感器(X,Y,Z):"+String.valueOf(values[0])+";"
                            +String.valueOf(values[1])+";"+String.valueOf(values[2]));
            }
        }
        @Override
        public void onAccuracyChanged(Sensor sensor, int accuracy) {
        }
}
```

由于Android Studio自带的模拟器不支持传感器，因此建议使用真机进行测试，连接手机测试代码的方式为：打开手机"设置"→"更多设置"→"开发者选项"→"开启开发者选项"→"允许USB调试"，然后将手机与电脑用USB线连接，在Android Studio中运行程序时选择该手机，此时程序将在手机上运行。如果"更多设置"选项中没有"开发者选项"，可以进入"设置"→"关于手机"→找到版本号，连续多次点击版本号即可开启"开发者选项"。

在实际的移动设备上，不一定所有的传感器都被支持，可以通过SensorManager对象的getSensorList()方法获取当前支持的传感器列表，以便查看该设备支持哪些传感器，示例代码如下：

```java
List<Sensor> sensors = sensorManager.getSensorList(Sensor.TYPE_ALL);
for (Sensor sensor:list){
    Log.i("传感器名称:******",sensor.getName());
}
```

# 附　录

## 附录 1　Android Studio 下载与安装

（1）下载地址：http://www.android-studio.org/。当前首页上默认的下载版本即为 3.2.0，可以直接点击"下载 ANDROID STUDIO"图标按钮或者下载列表中的"android-studio-ide-181.5014246-windows.exe"下载相应的 Windows 版本软件。如果版本有更新，可以在菜单栏"下载"中寻找版本号为 3.2.0 的历史版本。如附图 1.1 所示。

（2）下载完成后，双击 Android studio 安装文件"android-studio-ide-181.5014246-windows.exe"，进入安装界面，点击"Next"，开始安装。注意：默认情况下，Android studio 的安装目录是"C:\program Files\Android\Android studio"，Android SDK 的安装目录是一个隐藏目录"C:\Users\[当前用户]\AppData\Local\Android\Sdk"。如附图 1.2 所示。

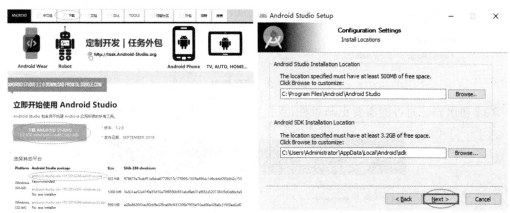

附图 1.1　安装包下载　　　　　　　附图 1.2　安装目录

（3）遇到无法访问 Android 插件列表时，点击"Cancel"取消即可，如附图 1.3 所示。

附图 1.3　无法访问 Android 插件

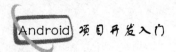

（4）组件下载完成后，会输出提示更新完成，此时点击"Finish"确认安装完成即可。

## 附录2  创建和运行第一个Android项目

### 1. 创建和运行第一个Android项目

（1）进入Android studio的开始页面，点击创建一个android项目，如附图2.1所示。

附图2.1  开始界面

（2）填入项目信息，第一个是应用名，第二个是公司域名（会自动生成包名），第三个是项目位置，填写完后点击下一步。例如填入项目名"HelloWorld"，公司域名（会自动生成包名）"turbosnail.cc"，项目路径"D:\AndroidStudioProjects\HelloWorld"，如附图2.2所示。

附图2.2  项目信息

（3）初次安装时可能需要更新组件，点击"Next"，等待更新，如附图2.3所示。

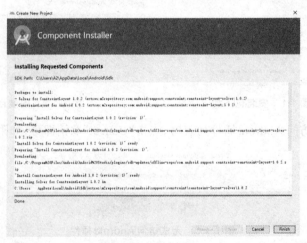

附图2.3  更新组件

（4）选择 Empty Activity，点击"Next"，如附图 2.4 所示。

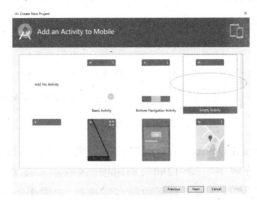

附图 2.4　添加项目

（5）显示默认的 Activity 信息，Activity 默认名称为：MainActivity，布局文件默认名称为：activity_main，不用修改。点击"Finish"，完成项目创建，如附图 2.5 所示。

附图 2.5　项目信息

（6）打开工作界面。注意，此时底部状态栏会显示 Gradle 下载更新，这一过程需要联网且耗费一些时间，直到显示类似"Gradle build finished in 6 s 350 ms（a minute ago）"的提示，才表示项目构建完成。Gradle 只是在第一次创建项目时下载更新，后续创建项目时不会再次耗时更新，如附图 2.6 所示。

附图 2.6　工作界面

## 2. 创建和启动手机模拟器

(1) 点击工具栏上的"AVD Manager"图标创建模拟器,如附图2.7所示。

附图2.7 模拟器

(2) 点击"Create Virtual Device"添加虚拟设备。然后选择手机型号,如选择"Nexus4",对应屏幕大小4.7英寸、分辨率"768×1280",如附图2.8所示。接着下一步,选择Android版本,默认选择"Nougat",也可以点击下载别的版本并选择,如附图2.9所示。

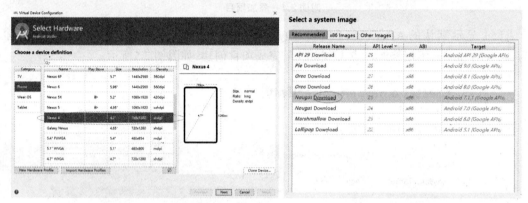

附图2.8 添加模拟器　　　　　　　　　　　附图2.9 选择版本

(3) 给模拟器命名,并可以进行一些设置。例如命名为"myphone",如附图2.10所示。创建完成后,点击图标启动模拟器,如附图2.11所示。

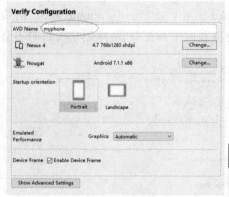

附图2.10 模拟器设置　　　　　　　　　　附图2.11 启动模拟器

(4) 启动后,可以看到模拟器,如附图2.12所示。

附图2.12 模拟器

3. 运行第一个 Android 项目

（1）点击 Android Stuido 工具栏上运行程序小图标，如附图2.13所示。

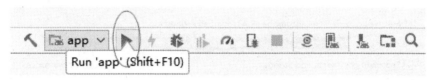

附图2.13 运行程序小图标

（2）选择模拟器。列表中会显示出所有用户创建出来的模拟器供选择，如果没有已经创建的模拟器，可以点击底部的"Create New Virtual Device"创建。勾选"Use same selection for future launches"可以使得后续程序均使用该模拟器，不用再进行模拟器选择，如附图2.14所示。

（3）程序运行后，模拟器上显示出"Hello World!"，效果如附图2.15所示。

附图2.14 选择模拟器　　　　附图2.15 运行结果

## 附录3　合理使用包(package)管理项目目录

包是Java提供的一种对项目代码分类管理的有效机制,它还可以起到访问控制、防止命名冲突等作用,在项目开发中对类、接口等按照一定的原则分类放在不同的包中,可以使得代码结构更加清晰、易于阅读和维护。

**[项目案例24]** 对AccountBook项目中的Java代码分包进行管理

为完成AccountBook项目,共编写Java类24个,分别包括Activity、Fragment……按照其功能,可以分别构建包来放置这些类。

(1) 打开AccountBook项目,在项目列表"java"→"cc.turbosnail.accountbook"上点击右键,选择"New"→"Package",输入包名fragment,点击"OK"。

(2) 将AccountAllFragment、AccountClothesFragment、AccountEatFragment等8个Fragment分别拖入项目目录下的fragment文件夹中,如附图3.1所示。

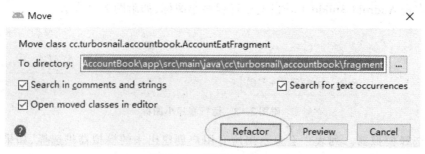

附图3.1　移动fragment

(3) 重复前两个步骤,创建包bean、dao、adapter、util,把对应代码拖入包文件夹中。原来的项目代码没有分包,阅读和查找均不方便,通过分包管理,代码的结构显得清新明了,便于维护。

在项目开发的过程中,建议提前规划包的结构,创建Java类时就直接划分到对应包中,养成良好的编码习惯,如附图3.2所示。

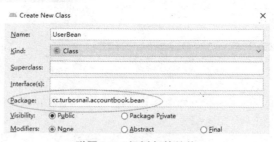

附图3.2　规划包的结构

(4) 分包前后的项目代码结构对比,如附图3.3所示。

# 附 录

附图3.3 分包前后项目结构对比

## 附录4　Android Studio常用设置

（1）自动导入包。"Alt+Enter"组合键可以自动导入Java包或者用户自定义的包到当前代码中。例如，在写Log后，点击Log，然后按下"Alt+Enter"组合键就可以选择并导入"import android.util.Log;"语句到代码中，如附图4.1所示。

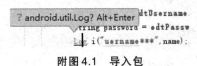

附图4.1　导入包

（2）跳转到类定义或者布局文件。"Ctrl+鼠标单击"可以从当前代码跳转并打开到类定义的源代码或者布局文件代码。例如，在setContentView（R.layout.activity_login）语句的activity_login上按住"Ctrl"并单击鼠标，就可以打开activity_login.xml文件。

（3）自动生成get/set访问器方法。在实体类中声明成员变量后，选择菜单栏"Code"→"Generate..."，就可以为成员变量选择自动生成set/get方法。

（4）自动格式化代码。写完Java或者XML代码后，选择菜单栏"Code"→"Reformate Code"，当前代码就会按照编码规范自动格式化。

（5）设置断点。在行标号和代码之间，紧靠行标号的位置单击，就会生成断点。断点可以用来跟踪测试代码的执行步骤，如附图4.2所示。

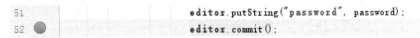

附图4.2　设置断点

使用断点时,通过Debug方式运行程序,如附图4.3所示,会打开Debug视图界面,此时可以看到断点出程序参数的值,还可手动选择继续向下执行代码。

附图4.3 Debug界面

(6)切换界面主题。菜单栏"File"→"Settings..."→"Appearance"→"Theme",选择Darcula主题,然后点击底部的"Apply",如附图4.4所示。

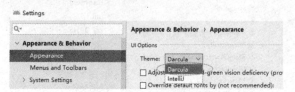

附图4.4 切换界面主题

此时,编辑器变成黑灰色主题,如附图4.5所示。

附图4.5 黑灰色主题